Bernhard Weigand

Analytical Methods for
Heat Transfer and Fluid Flow Problems

Bernhard Weigand

Analytical Methods for Heat Transfer and Fluid Flow Problems

With 80 Figures

 Springer

Professor Dr.-Ing. Bernhard Weigand
Universität Stuttgart
Institut für Thermodynamik der Luft- und Raumfahrt
Pfaffenwaldring 31
70569 Stuttgart
www.uni-stuttgart.de/itlr/
bw@itlr.uni-stuttgart.de

ISBN 3-540-22247-2 Springer Berlin Heidelberg New York

Library of Congress Control Number: 2004107594

Springer. Part of Springer Science+Business Media
springeronline.com

© Springer-Verlag Berlin Heidelberg 2004
Printed in Germany

Typesetting: data delivered by author
Cover design: deblik Berlin
Printed on acid free paper   62/3020/M - 5 4 3 2 1 0

For Irmi, my wife

Preface

Partial differential equations are the basis for nearly all technical processes in heat transfer and fluid mechanics. In my lectures over the past seven years I became aware of the fact that a lot of the students studying mechanical or aerospace engineering and also a lot of the engineers in industry today focus more and more on numerical methods for solving these partial differential equations. Analytical methods, taught in undergraduate mathematics, in thermodynamics and fluid mechanics, are quickly discarded, because most people believe that almost all problems, appearing in *real* applications, can easily be solved by numerical methods. In addition, most of the examples shown in basic lectures are *so simple* that the students develop the impression that analytical methods are inappropriate for more complicated *realistic technical* problems.

It was exactly the above described mind set, which inspired me several years ago to give lectures on analytical methods for heat and mass transfer problems. The basic idea of these lectures is to show some selected analytical methods and to explain their application to more complicated problems, which are technically relevant. Of course, this means that some of the standard analytical methods, might not be discussed in these lectures and are also not present in this book (for example integral transforms). On the other hand, it can be shown that the analytical methods discussed here are applicable to interesting problems and the student or engineer learns how to solve useful technical problems analytically.

This means that the main intent of this book is to show the usefulness of analytical methods, in a world, which focuses more and more on numerical methods. Of course, there is no doubt that the knowledge of numerical solution methods is very important and there is a big chance in using numerical tools to gain inside into flow physics and heat transfer characteristics. However, numerical methods are always dependent on grid quality and grid size and also on a lot of implementation features. Analytical methods can be used to validate and improve numerical methods. So the engineer might simplify a problem up to the extent that he can obtain an analytical solution. This analytical solution might be used afterwards to check and to improve the numerical solution for the full problem without any simplifications.

This book has been written for graduate students and engineers. The mathematics needed to understand the solution approach is developed mostly during the actual solution of the problem under consideration. This means that the book includes only few proofs. The reader is referred in these situations to other books for these more basic mathematical considerations. This approach has been taken in order to keep the focus of the book on the solution method itself and not to disrupt the analysis of the technical problem.

The book is structured in six chapters. Chapter 1 provides a short introduction to the topic.

Chapter 2 provides an introduction to the solution of linear partial differential equations. After discussing the classification and the character of the solutions of second-order partial differential equations, the method of separation of variables is discussed in detail.

Chapter 3 is concerned with the solution of thermal entrance problems for pipe and channel flows. This means that solutions of the energy equation are considered for a hydrodynamically fully developed velocity profile. These problems lead to the solution of Sturm-Liouville eigenvalue problems, which are discussed for laminar and turbulent flows and for different wall boundary conditions. For the problems considered in Chap. 3, axial heat conduction within the flow can be ignored, because the Peclet number in the flow is sufficiently large. This means that the problems under consideration are parabolic in nature. Because this sort of eigenvalue problems normally cannot be solved analytically, a numerical procedure is discussed on how to solve them. In Appendix C, this numerical solution method is explained in detail and the reader is provided links to an internet page, containing several source codes and executables.

Chapter 4 explains analytical solution methods for Sturm-Liouville eigenvalue problems for large eigenvalues. Here the focus is to explain an asymptotic analysis for a complicated problem, which is technically relevant. This chapter also provides comparisons between numerically and analytically predicted eigenvalues and constants. These comparisons show the usefulness of the analytical solution. Furthermore, it is explained how the method can be used for related problems.

In Chapter 5 the heat transfer in pipe and channel flows for small Peclet numbers is considered. In contrast to the problems discussed in Chaps. 3-4, the axial heat conduction in the fluid cannot be ignored. This leads to elliptic problems. A method is presented, which gives rise to solutions, which are as simple as the ones presented in Chap. 3. The extension of this method to more complicated problems, for example for the heat transfer in hydrodynamically fully developed duct flows with a heated zone of finite length, is also explained.

Chapter 6 is devoted finally to the solution of nonlinear partial differential equations. The idea behind this chapter was to provide a short overview about different solution methods for nonlinear partial differential equations. However, the main focus is on the derivation of similarity solutions. Here, different solution methods are explained. These are the method of dimensional analysis, group-theory methods and the method of the free parameter. The methods are demonstrated for a simple heat conduction problem as well as for complicated boundary layer problems.

Many people helped me in all phases of the preparation of this book. I am very grateful for many helpful discussions with my colleague Prof. Jens von Wolfersdorf concerning all aspects of the analytical solution methods. I also thank very much Martin Stricker and Marco Schüler who helped me with the figures. Many thanks also go to Dr. Grazia Lamanna for the helpful discussions and her support finishing this book. Also, I would like to thank Karl Straub very much for reading the manuscript.

I kindly acknowledge the permission of the ASME for reprinting the Figs. 1.2-1.3 and of ELSEVIER for reprinting the Figs. 3.7, 3.13-3.15, 5.1, 5.9-5.13, 5.18-5.19, 5.25 in this book. I also kindly acknowledge the permission of the Council of Mechanical Engineers (IMECHE) for reprinting Fig. 3.12, of KLUWER for reprinting the Figs. 5.3-5.4, B2-B7 and of SPRINGER for reprinting the Figs. 3.8, 3.9, 3.17-3.19 in this book. In addition, I kindly acknowledge the permission of Prof. E. Papoutsakis for reprinting the Figs. 5.3-5.4 and of Prof. Osterkamp, Dr. Zhang and Dr. Gosink for reprinting Fig. 1.4 in this book. The reference of the paper, where the figures have originally been published, is always included in the individual figure legend.

Finally, I am very grateful for the very good cooperation with Springer Press during the preparation of this manuscript. Here I would like to thank in particular Mrs. Maas, Mrs. Jantzen and Dr. Merkle for their support.

Filderstadt, April 2004   Bernhard Weigand

Contents

List of Symbols.. XV

List of Copyrighted Figures XIX

1 Introduction.. 1
    Problems.. 9

2 Linear Partial Differential Equations 11
    2.1 Classification of Second-Order Partial Differential Equations 11
    2.2 Character of the Solutions for the Partial Differential Equations 19
        2.2.1 Parabolic Second-Order Equations 19
        2.2.2 Elliptic Second-Order Equations 21
        2.2.3 Hyperbolic Second-Order Equations 23
    2.3 Separation of Variables 24
        2.3.1 One-Dimensional Transient Heat Conduction in a Flat Plate............ 24
        2.3.2 Steady-State Heat Conduction in a Rectangular Plate...................... 31
        2.3.3 Separation of Variables for the General Case of a Linear Second-Order Partial Differential Equation.. 37
        Problems.. 39

3 Heat Transfer in Pipe and Channel Flows (Parabolic Problems) 43
    3.1 Heat Transfer in Pipe and Channel Flows with Constant Wall Temperature 43
        3.1.1 Velocity Distribution of Hydrodynamically Fully Developed Pipe and Channel Flows.. 44
        3.1.2 Thermal Entrance Solutions for Constant Wall Temperature 46
        Properties of the Sturm-Liouville System....................................... 50
    3.2 Thermal Entrance Solutions for an Arbitrary Wall Temperature Distribution 66
    3.3 Flow and Heat Transfer in Axially Rotating Pipes with Constant Wall Heat Flux 69
        3.3.1 Velocity Distribution for the Hydrodynamically Fully Developed Flow in an Axial Rotating Pipe.. 72
        3.3.2 Thermal Entrance Solution for Constant Wall Heat Flux................. 75
        Problems.. 83

4 Analytical Solutions for Sturm - Liouville Systems with Large Eigenvalues **87**
 4.1 Heat Transfer in Turbulent Pipe Flow with Constant Wall Temperature 103
 4.2 Heat Transfer in an Axially Rotating Pipe with Constant Wall
 Temperature 109
 4.3 Asymptotic Expressions for other Thermal Boundary Conditions 113
 Problems 115

**5 Heat Transfer in Duct Flows for Small Peclet Numbers
(Elliptic Problems)** **117**
 5.1 Heat Transfer for Constant Wall Temperatures for x ≤ 0 and x > 0 120
 5.1.1 Heat Transfer in Laminar Pipe and Channel Flows for Small Peclet
 Numbers 131
 5.1.2 Heat Transfer in Turbulent Pipe and Channel Flows for Small Peclet
 Numbers 138
 5.2 Heat transfer for Constant Wall Heat Flux for x ≤ 0 and x > 0 142
 5.2.1 Heat Transfer in Laminar Pipe and Channel Flows for Small Peclet
 Numbers 147
 5.2.2 Heat Transfer in Turbulent Pipe and Channel Flows for Small Peclet
 Numbers 150
 5.3 Results for Heating Sections with a Finite Length 153
 5.3.1 Piecewise Constant Wall Temperature 154
 5.3.1 Piecewise Constant Wall Heat Flux 157
 5.4 Application of the Solution Method to Related Problems 159
 Problems 161

6 Nonlinear Partial Differential Equations **165**
 6.1 The Method of Separation of Variables 165
 6.2 Transformations Resulting in Linear Partial Differential Equations 172
 6.3 Functional Relations Between Dependent Variables 174
 6.3.1 Incompressible Flow over a Heated Flat Plate 174
 6.3.2 Compressible Flow over a Flat, Heated Plate 177
 6.4 Similarity Solutions 179
 6.4.1 Similarity Solutions for a Transient Heat Conduction Problem 179
 6.4.2 Similarity Solutions of the Boundary Layer Equations for Laminar
 Free Convection Flow on a Vertical Flat Plate 186
 6.4.3 Similarity Solutions of the Compressible Boundary Layer
 Equations 192
 Problems 198

**Appendix A: The Fully Developed Velocity Profile for Turbulent Duct
Flows** **203**

**Appendix B: The Fully Developed Velocity Profile in an Axially Rotating
Pipe** **215**

Appendix C: A Numerical Solution Method for Eigenvalue Problems 227
   C.1 Numerical Tools 229

**Appendix D: Detailed Derivation of Certain Properties of the Method for
Solving the Extended Graetz Problems** 235
   D.1 Symmetry of the Matrix Operator $\underset{\sim}{L}$ 235
   D.2 The Eigenfunctions Constitute a Set of Orthogonal Functions 236
   D.3 A detailed Derivation of Eq. (5.31) and Eq. (5.61) 237
   D.4 Simplification of the Expression for the Temperature Distribution (for
   Constant Wall Temperature) 239
   D.5 Simplification of the Expression for the Temperature Distribution (for
   Constant Wall Heat Flux) 240
   D.6 The Vector Norm $\left\| \vec{\Phi}_j \right\|^2$ 243

References 247

Index 257

List of Symbols

a	[m^2/s]	heat diffusivity
a_1, a_2	[-]	functions
A	[m^2]	flow area
A_j	[-]	constants
Bi	[-]	Biot number
c	[m]	velocity of sound
c_f	[-]	friction factor
c_p	[J/(kg K)]	specific heat at constant pressure
C	[-]	Chapman-Rubesin parameter
D	[m]	hydraulic diameter
E	[-]	dimensionless energy flow
F	[-]	flow index (0 for planar channel, 1 for pipe)
F_x, F_y, F_z	[N]	forces
G	[-]	function
h	[W/(m^{2}K)]	heat transfer coefficient
h	[m]	half channel height of a planar channel
$J_i(s)$	[-]	Bessel function of order i
k	[W/(m K)]	thermal conductivity
k	[-]	transformed eigenvalue
K	[W/m^3]	sink intensity
L	[m]	length scale (h for planar channel, R for pipe)
L_{th}	[m]	thermal entrance length
$\underset{\sim}{L}[\ ]$	[-]	matrix operator
l	[m]	mixing length
$M[\ ]$	[-]	operator
Ma$_\infty$	[-]	Mach number
Nu$_L$	[-]	Nusselt number based on L
Nu$_\infty$	[-]	Nusselt number for the fully developed flow
N	[-]	rotation rate
n	[m]	coordinate orthogonal to the flow direction
p	[Pa]	pressure
Pr	[-]	Prandtl number
Pr$_t$	[-]	turbulent Prandtl number

Pe_L	[-]	Peclet number based on L
Pe_t	[-]	turbulent Peclet number
R	[m]	pipe radius
R	[J/(kg K)]	gas constant
Re_L	[-]	Reynolds number based on L
Ri	[-]	Richardson number
t	[s]	time
T	[K]	temperature
T'	[K]	temperature fluctuation
T_W	[K]	wall temperature
T_b	[K]	bulk–temperature
u_τ	[m/s]	shear velocity
U	[m]	wetted perimeter
$\bar{u}$	[m/s]	mean velocity
u, v, w	[m/s]	velocity components
u', v', w'	[m/s]	fluctuating velocity components
V	[-]	dimensionless velocity gradient at the wall
x, y, z	[m]	coordinates
y^+	[-]	wall coordinate
Z	[-]	modified rotation parameter

Greek letter symbols

β	[1/K]	volumetric coefficient of expansion
β_1, β_2	[-]	constants
Γ	[-]	gamma function
ε_m	[m²/s]	eddy diffusivity for momentum transfer
$\varepsilon_{hx}, \varepsilon_{hy}, \varepsilon_{hr}$	[m²/s]	eddy diffusivity for heat transfer
ξ, η	[-]	characteristic coordinates
η	[-]	similarity variable
ϑ	[-]	transformed eigenfunction
Θ	[-]	dimensionless temperature
Θ_b	[-]	dimensionless bulk-temperature
λ_j	[-]	eigenvalue
μ	[kg/(m s)]	dynamic viscosity
ν	[m²/s]	kinematic viscosity
ρ	[kg/m³]	density
τ	[-]	dimensionless time
ϕ	[-]	enthalpy function

Φ_j	[-]	eigenfunctions
Φ_{Dis}	[1/s^2]	dissipation function
Φ	[m^2/s]	velocity potential
Ψ	[m^2/s]	stream function
ω	[1/s]	angular velocity

Subscripts

0		refers to inlet conditions
C		centerline of the duct
∞		free stream conditions
~, +		dimensionless quantities

Definition of non-dimensional Numbers

$$\text{Bi} = \frac{hD}{k}$$

Biot number

$$C = \frac{\rho\mu}{\rho_\infty \mu_\infty}$$

Chapman-Rubesin parameter

$$N = \frac{\text{Re}_\varphi}{\text{Re}_D}$$

Rotation rate

$$\text{Nu}_L = \frac{hL}{k} = \frac{-\frac{\partial T}{\partial n}\Big|_W L}{T_W - T_b}$$

Nusselt number

$$\text{Ma}_\infty = \frac{u_\infty}{\sqrt{\kappa p / \rho}}$$

Mach number

$$\text{Pe}_L = \frac{\bar{u} L}{a} = \text{Re}_L \text{Pr}$$

Peclet number

$$\text{Pr} = \frac{\mu c_p}{k} = \frac{\nu}{a}$$

Prandtl number

$$\text{Re}_L = \frac{\bar{u} L}{\nu}$$

Reynolds number

$$\mathrm{Re}_\varphi = \frac{w_W D}{\nu}$$

Rotational Reynolds number

$$\mathrm{Re}_\tau = \frac{u_\tau L}{\nu}$$

Shear stress Reynolds number

$$\mathrm{Ri} = \frac{2\dfrac{w}{r}\dfrac{\partial}{\partial r}(w r)}{\left(\dfrac{\partial u}{\partial r}\right)^2 + \left(r\dfrac{\partial}{\partial r}\left(\dfrac{w}{r}\right)\right)^2}$$

Richardson number

$$Z = N\frac{\mathrm{Re}_D}{2\mathrm{Re}_\tau} = N / \sqrt{\frac{c_f}{8}}$$

Modified rotation parameter

List of Copyrighted Figures

I kindly acknowledge the permission of the ASME for reprinting the Figs. 1.2-1.3, which have been published first in Burow P, Weigand B (1990) One-dimensional heat conduction in a semi-infinite solid with the surface temperature a harmonic function of time: A simple approximate solution for the transient behavior. Journal of Heat Transfer 112: 1076 – 1079 (Fig. 1).

I kindly acknowledge the permission of ELSEVIER for reprinting the Figs. 3.7, 5.9 and 5.18 which have been published first in Weigand B, Ferguson JR, Crawford ME (1997a) An extended Kays and Cawford turbulent Prandtl number model, Int. J. Heat Mass Transfer, 40: 4191- 4196 (Figs. 3-4). In addition, I kindly acknowledge the permission of ELSEVIER for reprinting the Figs. 3.13-3.15, which have been first published in Reich G, Beer H (1989) Fluid flow and heat transfer in an axially rotating pipe-I. Effect of rotation on turbulent pipe flow, Int. J. Heat Mass Transfer 32: 551-562 (Figs. 1, 3-4) and I kindly acknowledge the permission of ELSEVIER for reprinting the Figs. 5.1, 5.10 - 5.11, which have been first published in Weigand B (1996) An exact analytical solution for the extended turbulent Graetz problem with Dirichlet wall boundary conditions for pipe and channel flows, Int. J. Heat Mass Transfer 39: 1625-1637 (Figs. 1, 5-6). In addition, I kindly acknowledge the permission of ELSEVIER for reprinting the Figs. 5.12-5.13, 5.19 which have been first published in Weigand B, Kanzamar M, Beer H (2001) The extended Graetz problem with piecewise constant wall heat flux for pipe and channel flows, Int. J. Heat Mass Transfer 44: 3941-3952 (Figs. 1, 9).

I also kindly acknowledge the permission of the Council of Mechanical Engineers (IMECHE) for reprinting Fig. 3.12, which has been first published in White A (1964) Flow of a fluid in an axially rotating pipe, Journal Mechanical Engineering Science 6: 47-52 (Figs.10-11).

I kindly acknowledge the permission of KLUWER for reprinting the Figs. 5.3-5.4, which have been first published in Papoutsakis E, Ramkrishna D, Lim H (1980) The extended Graetz problem with Dirichlet wall boundary conditions, Applied Scientific Research 36: 13-34 (Figs. 1-3) and I kindly acknowledge the permission of KLUWER for reprinting Figs. B2-B7, which have been first published in Weigand B, Beer H (1994) On the universality of the velocity profiles of a turbulent flow in an axially rotating pipe, Applied Scientific Research 52: 115-132 (Figs. 5-7, 9-10, 12).

I kindly acknowledge the permission of SPRINGER for reprinting the Figs. 3.8, 3.9, 3.17-3.19 which have been first been published in Weigand B, Beer H (1989) Wärmebertragung in einem axial rotierenden, durchströmten Rohr im Bereich des thermischen Einlaufs, Wärme- und Stoffübertragung 24: 191-202

(Figs. 4 - 6, 9) and I kindly acknowledge the permission of SPRINGER for reprinting Fig. 5.25 which has been first published in Weigand B, Wrona F (2003) The extended Graetz problem with piecewise constant wall heat flux for laminar and turbulent flows inside concentric annuli, Heat and Mass Transfer 39: 313-320 (Fig. 2).

In addition, I kindly acknowledge the permission of Prof. E. Papoutsakis for reprinting the Figs. 5.3-5.4, which have been first published in Papoutsakis E, Ramkrishna D, Lim H (1980) The extended Graetz problem with Dirichlet wall boundary conditions, Applied Scientific Research 36: 13-34 (Figs. 1-3). I also kindly acknowledge the permission of Prof. Osterkamp, Dr. Zhang and Dr. Gosink for reprinting Fig. 1.4, which has been first published in Zhang T, Osterkamp TE, Gosink JP (1991) A model for the thermal regime of Permafrost within the depth of annual temperature variations. Proc. 3rd Int. Symp. on Therm. Eng. Sci. for Cold Regions, Fairbanks, Alaska, USA, pp. 341 – 347 (Fig. 3).

1 Introduction

Fluid flow and heat transfer problems are present in all of our daily life. For example, if we walk along a river and look at the water flowing with high speed over the river-bed, we actually observe a fluid mechanics problem. If we put some sugar into our coffee and stir it, we are faced with a complicated heat and mass transfer problem. In particular, convective heat transfer problems are present everywhere in our world. Most of the problems encountered in fluid mechanics or heat transfer are described by partial differential equations. One good example of such equations are the Navier-Stokes equations and the energy equation for an incompressible flow with constant fluid properties. If we consider a three-dimensional, steady-state problem, these equations read

$$\rho\left(u\frac{\partial u}{\partial x}+v\frac{\partial u}{\partial y}+w\frac{\partial u}{\partial z}\right)=F_x-\frac{\partial p}{\partial x}+\mu\left(\frac{\partial^2 u}{\partial x^2}+\frac{\partial^2 u}{\partial y^2}+\frac{\partial^2 u}{\partial z^2}\right) \tag{1.1}$$

$$\rho\left(u\frac{\partial v}{\partial x}+v\frac{\partial v}{\partial y}+w\frac{\partial v}{\partial z}\right)=F_y-\frac{\partial p}{\partial y}+\mu\left(\frac{\partial^2 v}{\partial x^2}+\frac{\partial^2 v}{\partial y^2}+\frac{\partial^2 v}{\partial z^2}\right) \tag{1.2}$$

$$\rho\left(u\frac{\partial w}{\partial x}+v\frac{\partial w}{\partial y}+w\frac{\partial w}{\partial z}\right)=F_z-\frac{\partial p}{\partial z}+\mu\left(\frac{\partial^2 w}{\partial x^2}+\frac{\partial^2 w}{\partial y^2}+\frac{\partial^2 w}{\partial z^2}\right) \tag{1.3}$$

$$\rho c\left(u\frac{\partial T}{\partial x}+v\frac{\partial T}{\partial y}+w\frac{\partial T}{\partial z}\right)=\mu\,\Phi_{Dis}+k\left(\frac{\partial^2 T}{\partial x^2}+\frac{\partial^2 T}{\partial y^2}+\frac{\partial^2 T}{\partial z^2}\right) \tag{1.4}$$

where Φ_{Dis} denotes the dissipation function in the energy equation. This function is given by

$$\Phi_{Dis}=2\left[\left(\frac{\partial u}{\partial x}\right)^2+\left(\frac{\partial v}{\partial y}\right)^2+\left(\frac{\partial w}{\partial z}\right)^2\right]+\left(\frac{\partial v}{\partial x}+\frac{\partial u}{\partial y}\right)^2+ \tag{1.5}$$

$$+\left(\frac{\partial w}{\partial y}+\frac{\partial v}{\partial z}\right)^2+\left(\frac{\partial u}{\partial z}+\frac{\partial w}{\partial x}\right)^2$$

This set of equations is closed by adding the incompressible mass continuity equation

$$\frac{\partial u}{\partial x} + \frac{\partial v}{\partial y} + \frac{\partial w}{\partial z} = 0 \tag{1.6}$$

In the above equations, x, y, z denote the cartesian coordinates, p indicates the pressure, T the temperature, F_x, F_y and F_z are forces, and u, v, w are the velocity components in the x, y and z direction, respectively.

From the above equations it can be seen that fluid flow and heat transfer problems are described by a set of partial differential equations. In general, these complicated differential equations can only be solved numerically. However, analytical solutions for fluid mechanics or heat transfer problems can still play an important role in science and in engineering, even in the current age of supercomputers. This is because analytical solutions have the big advantage of showing directly which parameters influence the solution. This is illustrated by the following short example (see Fig. 1.1). We are interested in the answer to the question on how a periodic change of the surface temperature of the earth will influence the temperature in the soil. Here, we are primarily interested in the behavior for increasing x, i.e. the distance from the surface, and growing values of time. Since we are only interested in the temperature change in the radial direction, we can approximate the earth as a semi-infinite body. In addition, we might use a cartesian coordinate system to describe this problem, because the radius of curvature of the earth is very large compared to all other dimensions.

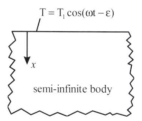

$T = T_1 \cos(\omega t - \varepsilon)$

x

semi-infinite body

Fig. 1.1: Semi-infinite body with a periodically changing surface temperature

If we write down the energy equation for this problem and if we consider the physical properties of the body to be constant, the energy equation can be simplified to

$$\rho c \frac{\partial T}{\partial t} = k \left(\frac{\partial^2 T}{\partial x^2} + \frac{\partial^2 T}{\partial y^2} + \frac{\partial^2 T}{\partial z^2} \right) \tag{1.7}$$

Furthermore, the heat conduction in the y- and z- direction can be neglected compared to the heat conduction in the x-direction. If we do so, we obtain

$$\frac{\partial T}{\partial t} = a \frac{\partial^2 T}{\partial x^2} \; ; \qquad a = \frac{k}{\rho c} \tag{1.8}$$

The boundary conditions for this problem are given by

$$x = 0: T = T_1 \cos\left(\omega t - \varepsilon\right) \tag{1.9}$$

$$t = 0: \ T = T_0 \ ; \quad x > 0 \tag{1.10}$$

where T_0 is the constant initial temperature of the solid body and ω and ε are given constants. The solution of this problem is (see Carslaw and Jaeger (1992))

$$\Theta = e^{-\eta} \cos\left(\tau - \eta - \varepsilon\right) + \frac{2}{\sqrt{\pi}} \int_0^{\eta/\sqrt{2\tau}} \cos\left[\tau - \frac{\eta^2}{2\mu^2} - \varepsilon\right] e^{-\mu^2} d\mu \tag{1.11}$$

with the dimensionless quantities

$$\Theta = \frac{T - T_0}{T_1} \ , \quad \tau = \omega t \ , \qquad \eta = x\sqrt{\frac{\omega}{2a}} \tag{1.12}$$

From the above given analytical solution (1.11), one sees very clearly that the solution consists of two parts. The second part of Eq. (1.11) determines the behavior of the solution for short times

$$\Theta_T = \frac{2}{\sqrt{\pi}} \int_0^{\eta/\sqrt{2\tau}} \cos\left[\tau - \frac{\eta^2}{2\mu^2} - \varepsilon\right] e^{-\mu^2} d\mu \tag{1.13}$$

For a fixed value of x ($\eta = $ const.), this part of the solution tends to zero with increasing time, as it can be seen from the upper boundary of the integral.

The first part of the solution (1.11) is a periodic part, which is multiplied by an exp-function

$$\Theta_P = e^{-\eta} \cos(\tau - \eta - \varepsilon) \tag{1.14}$$

From this part of the solution, we can see that the amplitude of the oscillation of the surface temperature decreases with increasing values of η. Additionally, one can notice that the maximum of the oscillation appears at different depths of the material with a time delay. Furthermore, it can be seen that the solution for Θ only depends on the following dimensionless quantities:

$$\tau = \omega t \tag{1.15}$$

$$\eta = x\sqrt{\frac{\omega}{2a}} \tag{1.16}$$

In this context, it is of special interest to have a closer look on the variable η. If $\sqrt{\omega/(2a)}$ is large (which means: ω is large or $a = k/(\rho c)$ is very small, for example because of a low heat conducting material), η will be large even for small values of x. This means that the temperature wave will be damped out very fast.

The behavior of the solution (1.11) is demonstrated for one example. We consider the case $\varepsilon = 0$. The complete solution is depicted in Fig. 1.2. It can be seen from Fig. 1.2 that the maximum temperature appears at different times for increasing values of η. In addition, it is clear from Fig. 1.2 that the solution satisfies the initial condition that $\Theta = 0$ for $\tau = 0$ inside the solid body (i.e. $\eta > 0$). Figure 1.3 shows only the first part of the solution Θ_p. As stated before, it can be seen that Θ_p is a very good approximation of the complete solution of the problem for larger times. For $\tau > 0.1$ the complete solution is nicely approximated.

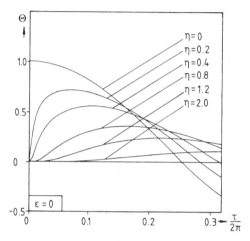

Fig. 1.2: Temperature distribution for $\varepsilon = 0$ as a function of η and τ (Burow and Weigand (1990))

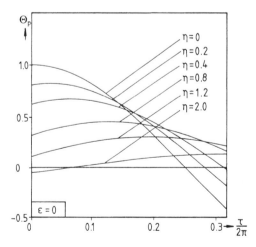

Fig. 1.3: Temperature distribution Θ_p for $\varepsilon = 0$ as a function of η and τ (Burow and Weigand (1990))

This also shows that, for some problems, it might suffice to determine the solution Θ_p and not the full solution of the problem. This sort of solution can be determined very easily also for more complicated variations of the surface temperature (see for example Myers (1987) or Carslaw and Jaeger (1992)). One example is depicted in Fig. 1.4, which shows the annual oscillation of the temperature in the soil in Alaska at various depths. The figure shows nicely how the minimum temperature at a certain depth in the soil gets shifted to different times. In addition it can be seen how the amplitude of the temperature oscillation decreases with increasing depth. This leads to the interesting fact, that at certain times in the year the temperature stratification in the soil can get reversed.

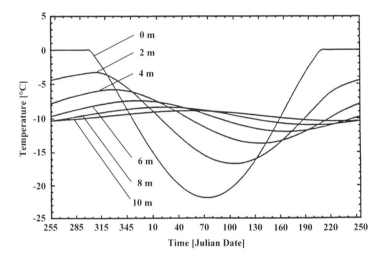

Fig. 1.4: Annual oscillation of the soil temperature in Alaska at different depths (Zhang et al. (1991))

The examples discussed above show very clearly where the strength of the analytical methods lay. The analytical solution shows clearly the dependence of the solution on the dimensional quantities. Furthermore, short time effects or long time behavior are clearly visible. Analytical solutions might therefore be derived for strongly simplified problems, before a numerical solution is carried out. The simplified model can then be used to understand the basic physical phenomena and behavior of the problem. Additionally, analytical solutions can be used for checking the numerical calculations and proving that all the settings, e.g. grid quality, meshing technique, numerical scheme and so on, are adequate for the considered problem. This is especially important for nonlinear problems, where grid size and grid quality might play a very important role for the final accuracy obtained by the numerical method.

In a next step, we want to address the question on how to classify the partial differential equations (PDE). In general, partial differential equations are classified by their order and by the fact if they are linear, quasilinear or nonlinear (the classification of second order linear partial differential equations is discussed in detail in Chap. 2).

An example of a linear second-order partial differential equation is the energy equation describing the heat conduction within a solid body, which has constant physical properties:

$$\rho c \frac{\partial T}{\partial t} = k \left(\frac{\partial^2 T}{\partial x^2} + \frac{\partial^2 T}{\partial y^2} + \frac{\partial^2 T}{\partial z^2} \right) \tag{1.17}$$

Another example of a linear second-order partial differential equation is the potential flow equation. Here we are considering an incompressible, irrotational flow with velocity components u and v. The velocity components are related to the velocity potential Φ by

$$u = \frac{\partial \Phi}{\partial x}, \qquad v = \frac{\partial \Phi}{\partial y} \tag{1.18}$$

From mass continuity, one obtains the following partial differential equation for the function Φ

$$\frac{\partial^2 \Phi}{\partial x^2} + \frac{\partial^2 \Phi}{\partial y^2} = 0 \tag{1.19}$$

In contrast to the above given equations, the partial differential equation, which describes the flow of a compressible inviscid flow, is a nonlinear second-order partial differential equation

$$\left(\left(\frac{\partial \Phi}{\partial x} \right)^2 - c^2 \right) \frac{\partial^2 \Phi}{\partial x^2} + 2 \frac{\partial \Phi}{\partial x} \frac{\partial \Phi}{\partial y} \frac{\partial^2 \Phi}{\partial x \partial y} + \left(\left(\frac{\partial \Phi}{\partial y} \right)^2 - c^2 \right) \frac{\partial^2 \Phi}{\partial y^2} = 0 \tag{1.20}$$

where Φ denotes in Eqs. (1.18–1.20) the velocity potential and c is the speed of sound. Another example of a nonlinear partial differential equation is the energy equation describing the one-dimensional transient heat conduction within a solid body whose thermal conductivity is a function of temperature. Additionally, an energy source term is incorporated in the material:

$$\rho c \frac{\partial T}{\partial t} = \left(\frac{\partial}{\partial x} \left(k(T) \frac{\partial T}{\partial x} \right) \right) + \dot{q}_i(x) \tag{1.21}$$

Summarizing, one can say that the order of a partial differential equation is given by the highest-ordered partial derivative appearing in the equation. A partial differential equation is called linear if it is linear in the unknown function and all its derivatives, i.e. the multiplying coefficients depend only on the independent variables. If the equation is linear in the highest-order derivative, it is called quasilinear. The Navier-Stokes equations are a good example of quasilinear equations. If the nonlinearity appears also in the highest-order derivative, it is called a nonlinear partial differential equation. Finally, it should be noted that the equation is called nonhomegeneous, if it contains one term which is only a function of the independent variables, otherwise it is called homogeneous. Equation (1.21) is a nonhomogeneous equation, because it contains a heat source which is a function of x. On the other hand Eqs. (1.17, 1.19, 1.20) are homogeneous equations.

The striking advantage in dealing with linear partial differential equations is that the final solution can be constructed by using the superposition method. This means that we can superimpose a number of solutions of much simpler problems to finally obtain the solution of a complicated problem. This is illustrated by the following simple example.

We consider the heat conduction within a flat plate which has the dimensions L in the x and y direction. The region contains a heat source and the four boundaries are kept at different temperatures. The physical properties of the material are considered constant. The problem is then described by the following sketch and equations:

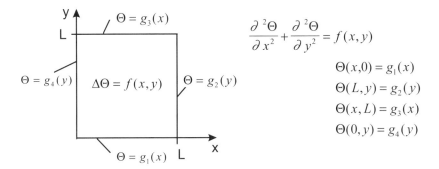

$$\frac{\partial^2 \Theta}{\partial x^2} + \frac{\partial^2 \Theta}{\partial y^2} = f(x,y)$$

$$\Theta(x,0) = g_1(x)$$
$$\Theta(L,y) = g_2(y)$$
$$\Theta(x,L) = g_3(x)$$
$$\Theta(0,y) = g_4(y)$$

Since the problem under consideration is described by a second-order linear partial differential equation, the solution can be obtained by superimposing the solutions of the following five simpler problems:

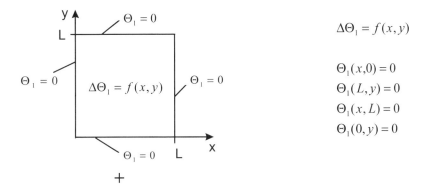

$$\Delta\Theta_1 = f(x,y)$$

$$\Theta_1(x,0) = 0$$
$$\Theta_1(L,y) = 0$$
$$\Theta_1(x,L) = 0$$
$$\Theta_1(0,y) = 0$$

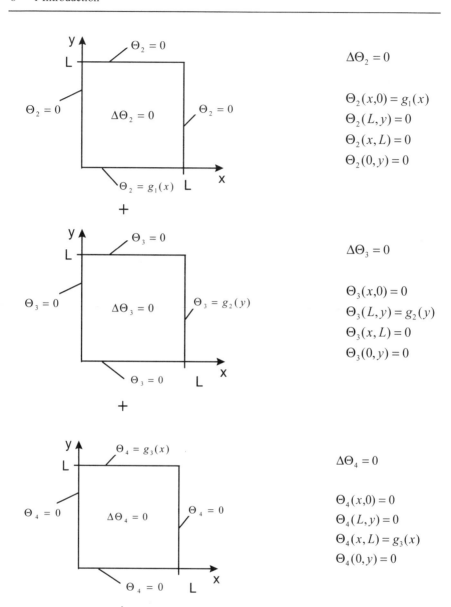

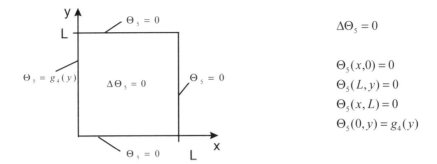

$$\Delta\Theta_5 = 0$$

$$\Theta_5(x,0) = 0$$
$$\Theta_5(L,y) = 0$$
$$\Theta_5(x,L) = 0$$
$$\Theta_5(0,y) = g_4(y)$$

After solving the above given individual problems, we know the solutions $\Theta_1, \Theta_2, \Theta_3, \Theta_4$ and Θ_5. The solution of the problem can then be obtained by simply adding all the solutions of the five sub-problems

$$\Theta = \sum_{i=1}^{5} \Theta_i \tag{1.22}$$

This can be visualized by looking at the individual equations and the related boundary conditions. Adding up the five individual partial differential equations results in

$$\frac{\partial^2 (\Theta_1 + \Theta_2 + \Theta_3 + \Theta_4 + \Theta_5)}{\partial x^2} + \frac{\partial^2 (\Theta_1 + \Theta_2 + \Theta_3 + \Theta_4 + \Theta_5)}{\partial y^2} = f(x,y) \tag{1.23}$$

By adding up the boundary conditions of the individual problems, the boundary conditions of the original problem will be obtained.

This example shows clearly how powerful the superposition method is. Furthermore, the superposition method makes linear partial differential equations so much easier to solve than nonlinear partial differential equations, where the superposition approach fails.

Problems

1-1. State for each of the following partial differential equations if it is linear, quasilinear or nonlinear. In addition, state if the equation is homogeneous or non-homogeneous and give its order:

$$\rho c \frac{\partial T}{\partial x} = k \frac{\partial^2 T}{\partial y^2}$$

$$\frac{\partial^2 T}{\partial x^2} + \frac{\partial^2 T}{\partial y^2} = f(x,y)$$

$$\frac{\partial^4 u}{\partial x^4} + \frac{\partial^4 u}{\partial y^4} + 3\frac{\partial^4 u}{\partial x^2 \partial y^2} + \frac{\partial u}{\partial x} = 5xy$$

$$2\frac{\partial^2 u}{\partial x^2} + \ln u = 3\left(x^2 + y^2\right)$$

$$\frac{\partial^2 u}{\partial x^2} + u\frac{\partial u}{\partial x} = e^y$$

1-2. Consider the steady-state temperature distribution in a quadratic plate. The problem is given by

$$\frac{\partial^2 \Theta}{\partial \tilde{x}^2} + \frac{\partial^2 \Theta}{\partial \tilde{y}^2} = 0$$

with the boundary conditions

$$\Theta(0, \tilde{y}) = 0$$
$$\Theta(1, \tilde{y}) = \sin(\pi\tilde{y})$$
$$\Theta(\tilde{x}, 0) = 0$$
$$\Theta(\tilde{x}, 1) = \sin(3\pi\tilde{x})$$

a.) Use the superposition method to solve this problem. Split therefore the problem in two different problems (one which includes only the non-homogeneous boundary condition $\Theta_1(1, \tilde{y}) = \sin(\pi\tilde{y})$ and one including the condition with $\Theta_2(\tilde{x}, 1) = \sin(3\pi\tilde{x})$).

b.) By adding the equations and boundary conditions of the two problems $\Theta_1 + \Theta_2$, show that the original problem is obtained.

c.) Show by inserting, that the two problems have the solutions:

$$\Theta_1 = \frac{\sinh(\pi\tilde{x})}{\sinh(\pi)}\sin(\pi\tilde{y})$$

$$\Theta_2 = \frac{\sinh(3\pi\tilde{y})}{\sinh(\pi)}\sin(3\pi\tilde{x})$$

d.)   Show by insertion that $\Theta = \Theta_1 + \Theta_2$ satisfies the original problem.

2 Linear Partial Differential Equations

Several important heat and fluid flow processes in technical applications and in nature can approximately be described by linear partial differential equations. As stated in the previous chapter, linear partial differential equations are normally much simpler to solve than nonlinear partial differential equations. In addition, a large body of literature exists on how to solve linear PDEs.

The following four chapters focus on the solution of linear partial differential equations. Chapter 2 is concerned with the classification of second order partial differential equations and presents a short introduction into the method of separation of variables. Chapter 3 focuses on convective heat transfer in laminar and turbulent pipe and channel flows. Here parabolic problems are considered and the general eigenvalue problems, associated with these equations, are explained. Chapter 4 discusses some specific methods for the analytical solution of eigenvalue problems, in the case of large eigenvalues. Finally, Chap. 5 deals with convective heat transfer problems in laminar or turbulent pipe and channel flows for low Peclet numbers (liquid metals). For this type of applications, the axial heat conduction within the flow can no longer be ignored and the resulting energy equation remains elliptic in nature. This has strong implications on the solution method for the energy equation.

2.1 Classification of Second-Order Partial Differential Equations

In the following, we are concerned with a linear second order partial differential equation which depends on the two independent variables x and y. The most general form of the homogeneous equation is given by

$$A(x,y)\frac{\partial^2 u}{\partial x^2} + 2B(x,y)\frac{\partial^2 u}{\partial x \partial y} + C(x,y)\frac{\partial^2 u}{\partial y^2} + \tag{2.1}$$

$$+ D(x,y)\frac{\partial u}{\partial x} + E(x,y)\frac{\partial u}{\partial y} + F(x,y)u = 0$$

where $A,...,F$ are constants or functions of x and y and are sufficiently differentiable in the domain of interest.

The form of Eq. (2.1) resembles the quadratic equation of a conic sections in analytical geometry. The equation

$$a x^2 + 2b\,xy + c\,y^2 + d\,x + e\,y + f = 0 \tag{2.2}$$

represents an ellipse, parabola or hyperbola depending on whether $ac\text{-}b^2 <, =, > 0$, respectively.

The classification of the second-order partial differential equation is based on the fact that Eq. (2.1) can be transformed into a standard form. This is very similar to the treatment of the quadratic equation (2.2) of conic sections in analytical geometry. We distinguish the following different cases (for the point x_0, y_0 under consideration):

$$1.\ B^2(x_0, y_0) - A(x_0, y_0)\, C(x_0, y_0) > 0 \tag{2.3}$$

hyperbolic type. There exist two real characteristic curves.

$$2.\ B^2(x_0, y_0) - A(x_0, y_0)\, C(x_0, y_0) = 0 \tag{2.4}$$

parabolic type. There exists one real characteristic curve.

$$3.\ B^2(x_0, y_0) - A(x_0, y_0)\, C(x_0, y_0) < 0 \tag{2.5}$$

elliptic type. The two characteristic curves are conjugate complex.

Each of these equations can be transformed into its standard form, if we introduce the following new coordinates into Eq. (2.1)

$$\xi = \xi(x, y) \tag{2.6}$$
$$\eta = \eta(x, y)$$

Is the equation of hyperbolic type, the standard form is given by

$$\frac{\partial^2 u}{\partial \xi \partial \eta} = H_1\!\left(\xi, \eta, u, \frac{\partial u}{\partial \xi}, \frac{\partial u}{\partial \eta}\right) \tag{2.7}$$

If one introduces the new coordinates

$$\bar{\xi} = \xi + \eta \tag{2.8}$$
$$\bar{\eta} = \xi - \eta$$

we obtain an alternative standard form of the hyperbolic equation

$$\frac{\partial^2 u}{\partial \bar{\xi}^2} - \frac{\partial^2 u}{\partial \bar{\eta}^2} = H_2\!\left(\bar{\xi}, \bar{\eta}, u, \frac{\partial u}{\partial \bar{\xi}}, \frac{\partial u}{\partial \bar{\eta}}\right) \tag{2.9}$$

If the equation is of parabolic type, then the standard form is given by

$$\frac{\partial^2 u}{\partial \xi^2} = H_3\left(\xi, \eta, u, \frac{\partial u}{\partial \xi}, \frac{\partial u}{\partial \eta}\right) \tag{2.10}$$

Finally, if the equation is of elliptic type, the standard form of the equation is given by

$$\frac{\partial^2 u}{\partial \tilde{\xi}^2} + \frac{\partial^2 u}{\partial \tilde{\eta}^2} = H_4\left(\tilde{\xi}, \tilde{\eta}, u, \frac{\partial u}{\partial \tilde{\xi}}, \frac{\partial u}{\partial \tilde{\eta}}\right) \tag{2.11}$$

where the new coordinates $\tilde{\xi}, \tilde{\eta}$ are defined by

$$\tilde{\xi} = \frac{1}{2}(\xi + \eta)$$
$$\tilde{\eta} = \frac{1}{2i}(\xi - \eta) \tag{2.12}$$

In order to obtain one of the above standard or canonical forms of the equation, we need to perform the coordinate transform given by Eq. (2.6). Since we want the transformed equation to be equivalent to the original equation, we assume that ξ and η are twice continuously differentiable and we insist that the Jacobian

$$\text{Jacobian} = \begin{vmatrix} \dfrac{\partial \xi}{\partial x} & \dfrac{\partial \xi}{\partial y} \\ \dfrac{\partial \eta}{\partial x} & \dfrac{\partial \eta}{\partial y} \end{vmatrix} = \frac{\partial \xi}{\partial x}\frac{\partial \eta}{\partial y} - \frac{\partial \xi}{\partial y}\frac{\partial \eta}{\partial x} \neq 0 \tag{2.13}$$

in the region under consideration. By assuming that Eq. (2.13) holds, we have always a unique transformation between x, y and ξ, η.

Use of the chain rule gives:

$$\left(\frac{\partial u}{\partial x}\right)_y = \left(\frac{\partial u}{\partial \xi}\right)_\eta \left(\frac{\partial \xi}{\partial x}\right)_y + \left(\frac{\partial u}{\partial \eta}\right)_\xi \left(\frac{\partial \eta}{\partial x}\right)_y \tag{2.14}$$

$$\left(\frac{\partial u}{\partial y}\right)_x = \left(\frac{\partial u}{\partial \xi}\right)_\eta \left(\frac{\partial \xi}{\partial y}\right)_x + \left(\frac{\partial u}{\partial \eta}\right)_\xi \left(\frac{\partial \eta}{\partial y}\right)_x$$

and also

$$\frac{\partial^2 u}{\partial x^2} = \frac{\partial^2 u}{\partial \xi^2}\left(\frac{\partial \xi}{\partial x}\right)^2 + 2\frac{\partial^2 u}{\partial \xi \partial \eta}\frac{\partial \xi}{\partial x}\frac{\partial \eta}{\partial x} + \frac{\partial^2 u}{\partial \eta^2}\left(\frac{\partial \eta}{\partial x}\right)^2 + \tag{2.15}$$
$$+ \frac{\partial u}{\partial \xi}\frac{\partial^2 \xi}{\partial x^2} + \frac{\partial u}{\partial \eta}\frac{\partial^2 \eta}{\partial x^2}$$

$$\frac{\partial^2 u}{\partial x \partial y} = \frac{\partial^2 u}{\partial \xi^2} \frac{\partial \xi}{\partial x} \frac{\partial \xi}{\partial y} + \frac{\partial^2 u}{\partial \xi \partial \eta} \left(\frac{\partial \xi}{\partial x} \frac{\partial \eta}{\partial y} + \frac{\partial \xi}{\partial y} \frac{\partial \eta}{\partial x} \right) +$$

$$+ \frac{\partial^2 u}{\partial \eta^2} \frac{\partial \eta}{\partial x} \frac{\partial \eta}{\partial y} + \frac{\partial u}{\partial \xi} \frac{\partial^2 \xi}{\partial x \partial y} + \frac{\partial u}{\partial \eta} \frac{\partial^2 \eta}{\partial x \partial y}$$

$$\frac{\partial^2 u}{\partial y^2} = \frac{\partial^2 u}{\partial \xi^2} \left(\frac{\partial \xi}{\partial y} \right)^2 + 2 \frac{\partial^2 u}{\partial \xi \partial \eta} \frac{\partial \xi}{\partial y} \frac{\partial \eta}{\partial y} + \frac{\partial^2 u}{\partial \eta^2} \left(\frac{\partial \eta}{\partial y} \right)^2 +$$

$$+ \frac{\partial u}{\partial \xi} \frac{\partial^2 \xi}{\partial y^2} + \frac{\partial u}{\partial \eta} \frac{\partial^2 \eta}{\partial y^2}$$

After inserting the above expressions into Eq. (2.1), one obtains

$$\bar{A}(\xi,\eta) \frac{\partial^2 u}{\partial \xi^2} + 2\bar{B}(\xi,\eta) \frac{\partial^2 u}{\partial \xi \partial \eta} + \bar{C}(\xi,\eta) \frac{\partial^2 u}{\partial \eta^2} + \tag{2.16}$$

$$\bar{D}(\xi,\eta) \frac{\partial u}{\partial \xi} + \bar{E}(\xi,\eta) \frac{\partial u}{\partial \eta} + F u = 0$$

with the coefficients

$$\bar{A} = A \left(\frac{\partial \xi}{\partial x} \right)^2 + 2B \frac{\partial \xi}{\partial x} \frac{\partial \xi}{\partial y} + C \left(\frac{\partial \xi}{\partial y} \right)^2 \tag{2.17}$$

$$\bar{B} = A \frac{\partial \xi}{\partial x} \frac{\partial \eta}{\partial x} + B \left(\frac{\partial \xi}{\partial x} \frac{\partial \eta}{\partial y} + \frac{\partial \xi}{\partial y} \frac{\partial \eta}{\partial x} \right) + C \frac{\partial \xi}{\partial y} \frac{\partial \eta}{\partial y}$$

$$\bar{C} = A \left(\frac{\partial \eta}{\partial x} \right)^2 + 2B \frac{\partial \eta}{\partial x} \frac{\partial \eta}{\partial y} + C \left(\frac{\partial \eta}{\partial y} \right)^2$$

$$\bar{D} = A \frac{\partial^2 \xi}{\partial x^2} + 2B \frac{\partial^2 \xi}{\partial x \partial y} + C \frac{\partial^2 \xi}{\partial y^2} + D \frac{\partial \xi}{\partial x} + E \frac{\partial \xi}{\partial y}$$

$$\bar{E} = A \frac{\partial^2 \eta}{\partial x^2} + 2B \frac{\partial^2 \eta}{\partial x \partial y} + C \frac{\partial^2 \eta}{\partial y^2} + D \frac{\partial \eta}{\partial x} + E \frac{\partial \eta}{\partial y}$$

Now we need to specify our change of variables, expressed by Eq. (2.6), in order to obtain one of the previously given standard or canonical forms. For example, if we want to obtain the hyperbolic equation in the form of Eq. (2.7) we have to assume that $\bar{A}$ and $\bar{C}$ are equal to zero. Because these two expressions are identical if we exchange ξ and η, we can achieve the condition $\bar{A} = \bar{C} = 0$, only when ξ and η are both solutions of the following equation

$$A\left(\frac{\partial\Omega}{\partial x}\right)^2 + 2B\left(\frac{\partial\Omega}{\partial x}\right)\left(\frac{\partial\Omega}{\partial y}\right) + C\left(\frac{\partial\Omega}{\partial y}\right)^2 = 0 \;\; ; \;\; \Omega{=}\xi \text{ or } \eta \tag{2.18}$$

Along a curve $\Omega = $ const. one has

$$d\Omega = \frac{\partial\Omega}{\partial x}dx + \frac{\partial\Omega}{\partial y}dy = 0 \quad => \frac{dy}{dx} = -\frac{\partial\Omega}{\partial x}\bigg/\frac{\partial\Omega}{\partial y} \tag{2.19}$$

we obtain from Eqs. (2.18) and (2.19) the following ordinary differential equation

$$A\left(\frac{dy}{dx}\right)^2 - 2B\left(\frac{dy}{dx}\right) + C = 0 \tag{2.20}$$

This differential equation can be solved for dy/dx and one gets the following two cases:

$$\frac{dy}{dx} = \left(B + \sqrt{B^2 - AC}\right)\bigg/A \tag{2.21}$$

$$\frac{dy}{dx} = \left(B - \sqrt{B^2 - AC}\right)\bigg/A$$

These two equations are known as the characteristic equations. They prescribe the functional relationship between the families of curves in the xy-plane for which ξ = const. and $\eta = $ const.. This means that a change of variables according to

$$\xi = f_1(x, y) \tag{2.22}$$
$$\eta = f_2(x, y)$$

will transform Eq. (2.1) into its standard form. From Eq. (2.21) it is apparent that there are three cases to be considered:

Case 1: Hyperbolic equation $(B^2 - AC) > 0$
The preceding analysis results in a canonical form for the hyperbolic equation. For this case we have two real characteristics, which can be obtained from the differential Eqs. (2.21).

Case 2: Elliptic equation $(B^2 - AC) < 0$
For this case, one obtains from Eq. (2.21) no real, but two conjugate complex solutions. Therefore, the elliptic equation has two conjugate complex characteristics. The elliptic case needs not to be recalculated again, because it can be deduced from the calculation of the hyperbolic case. This can be shown as follows: let us consider the canonical form

$$\frac{\partial^2 u}{\partial \xi \partial \eta} = H_1\left(\xi, \eta, u, \frac{\partial u}{\partial \xi}, \frac{\partial u}{\partial \eta}\right) \tag{2.7}$$

Now, we have two conjugate complex characteristics. Therefore, we can introduce into the above equation the coordinate transform given by Eq. (2.12)

$$\tilde{\xi} = \frac{1}{2}(\xi + \eta) \tag{2.12}$$

$$\tilde{\eta} = \frac{1}{2i}(\xi - \eta)$$

Use of the chain rule according to Eq. (2.15) gives

$$\frac{\partial^2 u}{\partial \xi \partial \eta} = \frac{\partial^2 u}{\partial \tilde{\xi}^2}\frac{\partial \tilde{\xi}}{\partial \xi}\frac{\partial \tilde{\xi}}{\partial \eta} + \frac{\partial^2 u}{\partial \tilde{\xi} \partial \tilde{\eta}}\left(\frac{\partial \tilde{\xi}}{\partial \xi}\frac{\partial \tilde{\eta}}{\partial \eta} + \frac{\partial \tilde{\xi}}{\partial \eta}\frac{\partial \tilde{\eta}}{\partial \xi}\right) +$$

$$+ \frac{\partial^2 u}{\partial \tilde{\eta}^2}\frac{\partial \tilde{\eta}}{\partial \xi}\frac{\partial \tilde{\eta}}{\partial \eta} + \frac{\partial u}{\partial \tilde{\xi}}\frac{\partial^2 \tilde{\xi}}{\partial \xi \partial \eta} + \frac{\partial u}{\partial \tilde{\eta}}\frac{\partial^2 \tilde{\eta}}{\partial \xi \partial \eta}$$

$$= \frac{\partial^2 u}{\partial \tilde{\xi}^2}\frac{1}{4} + \frac{\partial^2 u}{\partial \tilde{\xi} \partial \tilde{\eta}}\left(-\frac{1}{4i} + \frac{1}{4i}\right) + \frac{\partial^2 u}{\partial \tilde{\eta}^2}\frac{1}{4} = \frac{1}{4}\left(\frac{\partial^2 u}{\partial \tilde{\xi}^2} + \frac{\partial^2 u}{\partial \tilde{\eta}^2}\right)$$

and Eq. (2.7) transforms into the standard form for the elliptic type, given by Eq. (2.11)

$$\frac{\partial^2 u}{\partial \tilde{\xi}^2} + \frac{\partial^2 u}{\partial \tilde{\eta}^2} = H_4\left(\tilde{\xi}, \tilde{\eta}, u, \frac{\partial u}{\partial \tilde{\xi}}, \frac{\partial u}{\partial \tilde{\eta}}\right) \tag{2.11}$$

Case 3: Parabolic equation $(B^2 - AC) = 0$

For this case, it can be seen from Eq. (2.21) that only one family of real characteristics exists. Because of this fact, we can set in Eq. (2.17) for example $\eta = x$. Note that this is only possible if ξ depends on y, so that the Jacobian, defined by Eq. (2.13), is not zero. Then we obtain immediately from Eq. (2.17) that $\bar{C} = 0$ while $\bar{B}$ is equal to:

$$\bar{B} = A\frac{\partial \xi}{\partial x} + B\frac{\partial \xi}{\partial y} \tag{2.23}$$

This expression is identical to zero, as it can be deduced from Eq. (2.18) rewritten as follows - for the case $(B^2 - AC) = 0$ -

$$A\left(\frac{\partial \xi}{\partial x}\right)^2 + 2B\left(\frac{\partial \xi}{\partial x}\right)\left(\frac{\partial \xi}{\partial y}\right) + \frac{B^2}{A}\left(\frac{\partial \xi}{\partial y}\right)^2 = \frac{1}{A}\left(A\frac{\partial \xi}{\partial x} + B\frac{\partial \xi}{\partial y}\right)^2 = 0 \tag{2.24}$$

Finally, the standard form of the parabolic equation, given by Eq. (2.10), is obtained if we divide Eq. (2.16) by $\overline{A}$.

In order to illustrate the above shown classification, we will investigate some simple examples:

Example 1: The equation

$$\frac{\partial^2 u}{\partial x^2} + 2\frac{\partial^2 u}{\partial x \partial y} + \frac{\partial^2 u}{\partial y^2} = 0 \tag{2.25}$$

is parabolic, because the expression $B^2 - AC = 1 - 1 = 0$. The characteristic equation (2.21) reduces to

$$\frac{dy}{dx} = 1 \tag{2.26}$$

which has the solution

$$y - x = C_1 \tag{2.27}$$

If we make now a change of coordinates according to

$$\xi = y - x \tag{2.28}$$
$$\eta = x$$

where η has been selected arbitrarily, although always respecting the condition that the Jacobian, defined in Eq. (2.13), is not equal to zero. Introducing the new coordinates into Eq. (2.25), we obtain

$$\frac{\partial^2 u}{\partial \eta^2} = 0 \tag{2.29}$$

This equation has the general solution

$$u = \eta F(\xi) + H(\xi) = x F(y - x) + H(y - x) \tag{2.30}$$

Example 2: Consider the equation

$$y^4 \frac{\partial^2 u}{\partial x^2} - x^4 \frac{\partial^2 u}{\partial y^2} - 2yx^2 \frac{\partial u}{\partial x} = 0 \tag{2.31}$$

From this equation, we obtain $B^2 - AC = x^4 y^4 > 0$. This shows that Eq. (2.31) is hyperbolic everywhere except along the coordinate axis ($x = 0$ or $y = 0$).

From Eq. (2.21) we obtain the following two ordinary differential equations for the characteristics

$$\frac{dy}{dx} = \left(B + \sqrt{B^2 - AC}\right)\bigg/A = \frac{x^2 y^2}{y^4} = \left(\frac{x}{y}\right)^2 \tag{2.32}$$

and

$$\frac{dy}{dx} = -\left(\frac{x}{y}\right)^2 \tag{2.33}$$

From the two Eqs. (2.32 - 2.33) we calculate the characteristic curves to be:

$$\frac{1}{3}\left(y^3 - x^3\right) = C_1 \quad ; \qquad \frac{1}{3}\left(y^3 + x^3\right) = C_2 \tag{2.34}$$

In order to transform Eq. (2.31) into its standard form, we have to perform the following coordinate transformation

$$\xi = \frac{1}{3}\left(y^3 - x^3\right) \tag{2.35}$$

$$\eta = \frac{1}{3}\left(y^3 + x^3\right)$$

Using Eqs. (2.14-2.15), one obtains

$$-4x^4 y^4 \frac{\partial^2 u}{\partial \xi \partial \eta} - 2\left(xy^4 + yx^4\right)\frac{\partial u}{\partial \xi} + 2\left(xy^4 - yx^4\right)\frac{\partial u}{\partial \eta} \tag{2.36}$$

$$-2yx^2\left(-x^2 \frac{\partial u}{\partial \xi} + x^2 \frac{\partial u}{\partial \eta}\right) = 0$$

and after simplifying and replacing x,y by ξ and η we finally get

$$\frac{\partial^2 u}{\partial \xi \partial \eta} - \frac{1}{3}\frac{1}{\xi - \eta}\frac{\partial u}{\partial \xi} - \frac{1}{3}\frac{3\xi - \eta}{\eta^2 - \xi^2}\frac{\partial u}{\partial \eta} = 0 \tag{2.37}$$

Example 3: We consider the wave equation

$$\frac{\partial^2 \Phi}{\partial t^2} - c^2 \frac{\partial^2 \Phi}{\partial x^2} = 0 \tag{2.38}$$

This equation describes for example the one-dimensional propagation of sound in a pipe. Because $B^2 - AC > 0$, the equation is hyperbolic in the region of interest. The characteristic equations are given by

$$\frac{dt}{dx} = \left(B + \sqrt{B^2 - 4AC}\right)\bigg/2A = \frac{2c}{-2c^2} = -\frac{1}{c} \tag{2.39}$$

and

$$\frac{dt}{dx} = \frac{1}{c} \tag{2.40}$$

From these two equations we obtain $ct + x = C_1$ and $-ct + x = C_2$ and we get the following new coordinates:

$$\xi = x + ct \tag{2.41}$$
$$\eta = x - ct$$

If we introduce this new coordinates into Eq. (2.38), we get the following simple partial differential equation

$$\frac{\partial^2 \Phi}{\partial \xi \partial \eta} = 0 \tag{2.42}$$

which has the general solution

$$\Phi(\xi, \eta) = \Psi_1(\eta) + \Psi_2(\xi) \tag{2.43}$$

or rewritten in x, t coordinates

$$\Phi(x, t) = \Psi_1(x - ct) + \Psi_2(x + ct) \tag{2.44}$$

This shows that the solution of Eq. (2.38) can be expressed as the superposition of two waves, which move with constant velocity c into different directions of the solution domain. This shows also nicely how the information in the problem is transferred by the two real characteristics. The solution obtained here is known in literature as the d'Alembert solution.

2.2 Character of the Solutions for the Partial Differential Equations

In the preceding section we have concentrated on the classification of the different second-order partial differential equations. However, for solving actual physical problems, it is of great importance to discuss also the character of the solutions and the associated boundary conditions.

2.2.1 Parabolic Second-Order Equations

Let us start our discussion with the parabolic partial differential equation. As stated before, this equation has one real characteristic. As an example, we look at the heat conduction equation for a one-dimensional unsteady conduction problem in a slab (see Fig. 2.1). The slab has the thickness l and the spatial coordinate ranges from $0 < x < l$.

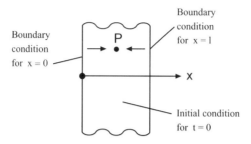

Fig. 2.1: Transient heat conduction in a slab

Assuming that the material properties of the slab are constant, the temperature distribution in the slab can be obtained from the solution of the following equation

$$\rho c \frac{\partial T}{\partial t} = k \frac{\partial^2 T}{\partial x^2} \tag{2.45}$$

The temperature distribution of the slab at the beginning of the process ($t = 0$) is given. This is the initial condition of the problem

$$T(0,x) = f_1(x) \tag{2.46}$$

In addition to this initial condition, boundary conditions have to be prescribed at the surface of the slab for $x = 0$ and $x = l$. Here the following different types of boundary conditions are possible:

- Boundary conditions of the first kind (Dirichlet wall boundary conditions). Here the temperature at the boundary is specified, for example

$$T(t,0) = f_2(t), \qquad T(t,l) = f_3(t) \tag{2.47}$$

- Boundary conditions of the second kind (Neumann conditions). For this type of boundary conditions the gradient is specified at the boundaries, for example

$$\left(\frac{\partial T}{\partial x}\right)_{x=0} = f_4(t), \qquad \left(\frac{\partial T}{\partial x}\right)_{x=l} = f_5(t) \tag{2.48}$$

- Boundary conditions of the third kind. Here a combination of temperature and gradient is prescribed at the surface. Such boundary conditions are relatively common in technical systems, since they describe, for example, the case of a slab which is heated or cooled by a fluid at temperature T_1 or T_2 flowing over the boundaries of the slab. A typical example is

$$k \left(\frac{\partial T}{\partial x} \right)_{x=0} = h_1 \left(T\left(t,0\right) - T_1 \right), \tag{2.49}$$

$$k \left(\frac{\partial T}{\partial x} \right)_{x=l} = h_2 \left(T\left(t,l\right) - T_2 \right)$$

where h_1 and h_2 are heat transfer coefficients.

Of course, all the above mentioned boundary conditions can be present in any possible combination, for example: at $x = 0$, a constant wall temperature is prescribed; whereas, at $x = l$, a boundary condition of the third kind is applied.

Summarising the above discussion, it can be seen that for the parabolic second-order partial differential equation an initial condition together with boundary conditions at the surface need to be specified. This means that the temperature at an arbitrary point P in the domain (see Fig. 2.1) is always influenced by the wall boundary conditions at $x = 0$ and $x = l$. In addition, all disturbances, which are specified for $t = 0$, will propagate into the solution domain for all subsequent times. On the other hand disturbances, which are introduced at a later time t_1, can not influence the solution for $t < t_1$. This shows nicely the character of the solution which depends only on one real characteristic.

2.2.2 Elliptic Second-Order Equations

For the elliptic equation, the characteristic curves are families of conjugate complex functions. In order to investigate the character of the solutions and the boundary conditions needed for this type of equations, we select as an example the steady-state heat conduction in a square plate $\left(0 \leq x \leq a, \;\; 0 \leq y \leq a \right)$, as shown in Fig. 2.2.

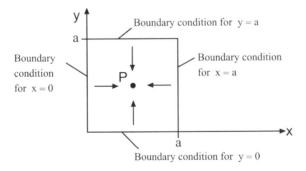

Fig. 2.2: Steady-state heat conduction in a square plate

The steady-state temperature distribution is obtained from the solution of the following equation:

$$\frac{\partial^2 T}{\partial x^2} + \frac{\partial^2 T}{\partial y^2} = 0 \qquad (2.50)$$

For this type of equation, boundary conditions have to be prescribed along each point of the boundary. We might illustrate this for the above given example. Here the following different types of boundary conditions can be assigned:

- Boundary conditions of the first kind (Dirichlet conditions). The temperature is specified at each point of the boundary

$$T(x,0) = T_1(x), \qquad T(x,a) = T_2(x)$$
$$T(0,y) = T_3(y), \qquad T(a,y) = T_4(y) \qquad (2.51)$$

- Boundary conditions of the second kind (Neumann conditions). The heat flux normal to the wall is specified along the boundary.

$$\left(\frac{\partial T}{\partial x}\right)_{x=0} = f_1(y), \qquad \left(\frac{\partial T}{\partial x}\right)_{x=a} = f_2(y)$$
$$\left(\frac{\partial T}{\partial y}\right)_{y=0} = f_3(x), \qquad \left(\frac{\partial T}{\partial y}\right)_{y=a} = f_4(x) \qquad (2.52)$$

- Boundary conditions of the third kind. This might be again a combination of the normal gradient at the surface and the temperature. One example is:

$$k\left(\frac{\partial T}{\partial x}\right)_{x=0} = f_1(T(0,y) - T_1), \quad k\left(\frac{\partial T}{\partial x}\right)_{x=a} = f_2(T(a,y) - T_2)$$
$$k\left(\frac{\partial T}{\partial y}\right)_{y=0} = f_3(T(x,0) - T_3), \quad k\left(\frac{\partial T}{\partial y}\right)_{y=a} = f_4(T(x,a) - T_4) \qquad (2.53)$$

Again, the boundary conditions can be applied as mixed boundary conditions.

From the above examples, it can be seen that for an elliptic equation we have to deal with a boundary-value problem, whereas for the parabolic equation, we had to solve a combined initial, boundary-value problem. This means that for the elliptic problem any disturbance, which is brought into the region of interest (for example by slightly changing one boundary condition) will immediately influence the solution of the problem at a given point in the domain (see point P in Fig. 2.2).

2.2.3 Hyperbolic Second-Order Equations

Hyperbolic partial differential equations mainly appear in vibration and wave problems. These equations have two real characteristics. The one-dimensional wave equation for a perfectly flexible string serves here as an example to explain the character of the solution and the associated boundary conditions (see Fig. 2.3).

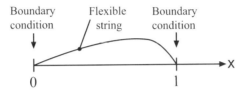

Fig. 2.3: Flexible string

The differential equation is given by

$$\frac{\partial^2 u}{\partial x^2} - \frac{1}{b^2}\frac{\partial^2 u}{\partial t^2} = 0 \tag{2.54}$$

where b is a constant. As shown in the previous section (see Eq. (2.44)), the general solution of this equation is given by

$$u(x,t) = \psi_1(x+bt) + \psi_2(x-bt) \tag{2.55}$$

Therefore, it would be easiest to prescribe boundary conditions for Eq. (2.54) along two parts of the characteristics intersecting at one point. This would be a complete initial value problem. However, for most physical problems, this description of the boundary conditions is not typical. Instead, the following boundary conditions might be normally applied:

- Boundary conditions of the first kind. Here the deflection of the string at the location $x = 0$ and $x = l$ might be prescribed.

$$u(0,t) = f_1(t), \qquad u(l,t) = f_2(t) \tag{2.56}$$

If the string is fixed at the locations $x = 0$ and $x = l$, f_1 and f_2 will be zero.

- Boundary conditions of the second kind. Here we will prescribe $\partial u/\partial x$ for $x = 0$ and $x = l$.

- For the boundary conditions of the third kind, we will specify a combination of u and $\partial u/\partial x$ for $x = 0$ and $x = l$.

As for all other types of equations, the boundary conditions can also be applied in a mixed form.

In addition to the two boundary conditions at $x = 0$ and $x = l$, initial conditions for $t = 0$ have to be specified for the problem. These initial conditions for the finite string could be that $u(x,0) = f_3(x)$, $\partial u / \partial t(x,0) = f_4(x)$.

2.3 Separation of Variables

This section is devoted to an introduction into the method of separation of variables. This method is one of the most commonly used methods for solving linear partial differential equations. The method is explained in the following sections by two basic examples. In the next chapters, more advanced problems are considered.

2.3.1 One-Dimensional Transient Heat Conduction in a Flat Plate

We consider the energy equation for heat conduction in a flat plate. The problem is depicted in Fig. 2.4. The plate has thickness δ and length L. At the two surfaces $x = 0$ and $x = \delta$ the plate is subjected to constant temperatures.

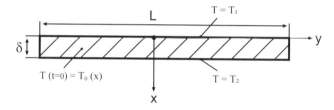

Fig. 2.4: Transient heat conduction in a flat plate

Under the assumption that the material properties of the plate are constant, the energy equation takes the following form.

$$\rho c \frac{\partial T}{\partial t} = k \left(\frac{\partial^2 T}{\partial x^2} + \frac{\partial^2 T}{\partial y^2} \right) \tag{2.57}$$

We now assume that the dimension L is much larger than δ, so that the heat conduction in the y-direction is negligible compared to the heat conduction in the x-direction. Therefore, the problem simplifies to

$$\frac{\partial T}{\partial t} = a \frac{\partial^2 T}{\partial x^2} \tag{2.58}$$

where $a = k/(\rho c)$ is the thermal diffusivity of the material. Eq. (2.57) has to be solved together with the following boundary conditions

$$x = 0 : T = T_1 \tag{2.59}$$
$$x = \delta : T = T_2$$

and the initial condition

$$t = 0 : T = T_0(x) \tag{2.60}$$

This problem is described by a parabolic equation. This means that one real characteristic exists for the solution. Before solving the above given problem, we first introduce the dimensionless quantities

$$\tilde{x} = \frac{x}{\delta}, \ \tilde{t} = \frac{at}{\delta^2}, \ \Theta = \frac{T - T_1}{T_2 - T_1} \tag{2.61}$$

This results in the following problem

$$\frac{\partial \Theta}{\partial \tilde{t}} = \frac{\partial^2 \Theta}{\partial \tilde{x}^2} \tag{2.62}$$

with the boundary conditions

$$\tilde{x} = 0 : \Theta = 0 \tag{2.63}$$
$$\tilde{x} = 1 : \Theta = 1$$
$$\tilde{t} = 0 : \Theta = (T_0(\tilde{x}) - T_1)/(T_2 - T_1) = \Theta_0(\tilde{x})$$

Before applying the method of separation of variables to the Eqs. (2.62 – 2.63), we want to investigate the solution domain, shown in Fig. 2.5. Here the parabolic nature of the problem is clearly visible. The initial condition for $\tilde{t} = 0$ is propagated into the solution domain for larger times. Any disturbance introduced into the problem at $\tilde{t} = \tilde{t}_1$ will therefore only influence the solution at subsequent times. The solution for $\tilde{t} < \tilde{t}_1$ stays unchanged.

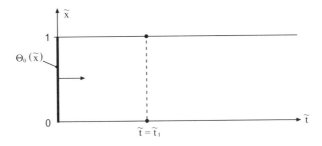

Fig. 2.5: Solution domain for the transient conduction in the plate

Let us assume that the solution of the problem can be expressed in the form

$$\Theta = H(\tilde{t})G(\tilde{x}) \tag{2.64}$$

Introducing Eq. (2.64) into the boundary conditions, results in:

$$\tilde{x} = 0 : H(\tilde{t})G(0) = 0 \tag{2.65}$$
$$\tilde{x} = 1 : H(\tilde{t})G(1) = 1$$

From this equation, we notice immediately that using the expression (2.64) can not result in a solution to the present problem, because the boundary condition for $\tilde{x} = 1$ can not be satisfied if $H(\tilde{t})$ is an arbitrary function of $\tilde{t}$. Therefore, we conclude that we first have to make the boundary conditions homogenous, in order to find a solution with the help of Eq. (2.64). This can be done by splitting the solution into two parts

$$\Theta = \Theta_S(\tilde{x}) + \Theta_T(\tilde{x},\tilde{t}) \tag{2.66}$$

The steady-state solution Θ_S is simply a linear distribution and is given by

$$\Theta_S(\tilde{x}) = \tilde{x} \tag{2.67}$$

Introducing Eq. (2.66) into the Eqs. (2.62-2.63) results in the following problem for Θ_T

$$\frac{\partial \Theta_T}{\partial \tilde{t}} = \frac{\partial^2 \Theta_T}{\partial \tilde{x}^2} \tag{2.68}$$

with the boundary conditions

$$\tilde{x} = 0 : \Theta_T = 0 \tag{2.69}$$
$$\tilde{x} = 1 : \Theta_T = 0$$
$$\tilde{t} = 0 : \Theta_T = \Theta_0(\tilde{x}) - \Theta_S(\tilde{t} = 0) = \Theta_0(\tilde{x}) - \tilde{x}$$

Introducing now again the product of functions, given by Eq. (2.64), we obtain the boundary conditions

$$\tilde{x} = 0 : H(\tilde{t})G(0) = 0 \qquad \Rightarrow \quad G(0) = 0 \tag{2.70}$$
$$\tilde{x} = 1 : H(\tilde{t})G(1) = 0 \qquad \Rightarrow \quad G(1) = 0$$

This shows that the expression given by Eq. (2.64) is able to satisfy the two boundary conditions of the problem. Therefore, this approach promises to be successful. Introducing Eq. (2.64) into the partial differential equation (2.68), one obtains

$$H'(\tilde{t})G(\tilde{x}) = G''(\tilde{x})H(\tilde{t}) \tag{2.71}$$

By separating the variables, this equation can be rewritten as

$$\frac{H'(\tilde{t})}{H(\tilde{t})} = \frac{G''(\tilde{x})}{G(\tilde{x})} \tag{2.72}$$

The left hand side of this equation is only a function of $\tilde{t}$, whereas the right hand side is only a function of $\tilde{x}$. Therefore, both sides of the equation must be constant. This constant is set to C_1.

$$\frac{H'(\tilde{t})}{H(\tilde{t})} = \frac{G''(\tilde{x})}{G(\tilde{x})} = C_1 \tag{2.73}$$

Let us first investigate the differential equation for the function H. This equation takes the form

$$\frac{H'(\tilde{t})}{H(\tilde{t})} = C_1 \tag{2.74}$$

and can easily been integrated to give

$$H(\tilde{t}) = C_2 \exp\left(C_1 \tilde{t}\right) \tag{2.75}$$

If we have a closer look at this equation, it can be seen that the function $H(\tilde{t})$ tends to infinity for $\tilde{t} \to \infty$ if $C_1 > 0$. However, this would not lead to a physically meaningful solution for the problem, because the temperature would tend to infinity for large times. For $C_1 = 0$, Eq. (2.75) results in a constant for $H(\tilde{t})$ and the time dependence of the solution would be lost. Therefore, we can conclude that the constant C_1 must always be smaller than zero for the problem under consideration. This can be expressed by replacing the constant by $C_1 = -\lambda^2$. Then we obtain for the function H

$$H(\tilde{t}) = C_2 \exp\left(-\lambda^2 \tilde{t}\right) \tag{2.76}$$

For the function $G(\tilde{x})$, one obtains the following ordinary differential equation from Eq. (2.73)

$$\frac{G''(\tilde{x})}{G(\tilde{x})} = -\lambda^2 \quad \Rightarrow \quad G'' + \lambda^2 G = 0 \tag{2.77}$$

This equation has to be solved together with the homogeneous boundary conditions given by Eq. (2.70). It has the trivial solution $G = 0$ and will have further solutions for selected values of λ. These selected values of λ are called the eigenvalues of Eq. (2.77). The problem given by Eq. (2.77) and associated boundary conditions, see Eq. (2.70), is called an eigenvalue problem. This sort of problem is discussed in more details in Chap. 3. The general solution of Eq. (2.77) is given by

$$G(\tilde{x}) = C_3 \sin(\lambda \tilde{x}) + C_4 \cos(\lambda \tilde{x}) \tag{2.78}$$

where the solution must satisfy the two boundary conditions

$$G(0) = 0, \quad G(1) = 0 \tag{2.79}$$

From the boundary condition $G(0) = 0$, it follows that C_4 is zero and one obtains

$$G = C_3 \sin(\lambda \tilde{x}) \tag{2.80}$$

Now, if we apply the second boundary condition $G(1) = 0$, we find

$$0 = C_3 \sin(\lambda \cdot 1) \tag{2.81}$$

Eq. (2.81) shows that either C_3 has to be zero (which would be the trivial solution of the problem, where $G = 0$) or that $\sin(\lambda)$ has to be zero. The latter is only possible if

$$\lambda = n\pi \qquad \text{with} \quad n=1,2,3,... \tag{2.82}$$

These special values of λ are the eigenvalues of Eq. (2.77) and are shown in Fig. 2.6.

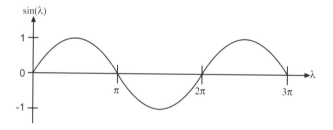

Fig. 2.6: Eigenvalues for the eigenfunction $\sin(\lambda)$

The solution for Θ_T is obtained from the Eqs. (2.75) and (2.80) as

$$\Theta_T = C_2 \exp\left(-\lambda^2 \tilde{t}\right) C_3 \sin(\lambda \tilde{x}) \tag{2.83}$$

For simplicity, we combine the constants C_2 and C_3 and have

$$\Theta_T = C \sin(\lambda \tilde{x}) \exp\left(-\lambda^2 \tilde{t}\right) \tag{2.84}$$

Now we try to fulfill the initial condition using Eq. (2.84). Inserting Eq. (2.84) into Eq. (2.69) results in

$$\tilde{t} = 0 : \Theta_T = \Theta_0\left(\tilde{x}\right) - \tilde{x} = C \sin\left(\lambda \tilde{x}\right) \exp\left(-\lambda^2 \, 0\right) \tag{2.85}$$
$$\Rightarrow \Theta_0\left(\tilde{x}\right) - \tilde{x} = C \sin\left(\lambda \tilde{x}\right)$$

From the equation above, one can see that Eq. (2.84) is automatically a solution of the problem if

$$\Theta_0\left(\tilde{x}\right) = \tilde{x} + C\sin\left(n\pi x\right), \quad n = 1, 2, 3, \ldots \tag{2.86}$$

However, from Eq. (2.82) it is clear that there is an infinite number of eigenvalues. Because the partial differential equation is linear, we use the principle of superposition to construct the final solution of the problem. This means that the solution will be given by

$$\Theta_T = \sum_{n=1}^{\infty} C_n \sin(\lambda_n\tilde{x})\exp\left(-\lambda_n^2\tilde{t}\right) \tag{2.87}$$

This solution has to fulfil the initial condition. Inserting Eq. (2.87) into Eq. (2.69) results in

$$\Theta_0(\tilde{x}) - \tilde{x} = \sum_{n=1}^{\infty} C_n \sin(\lambda_n\tilde{x}) \tag{2.88}$$

which means that we have to represent the function $\Theta_0(\tilde{x}) - \tilde{x}$ by a Fourier series (see Stephenson (1986), Myint-U and Debnath (1987), Zauderer (1989) or Sommerfeld (1978)).

In order to obtain the unknown coefficients C_n, from Eq. (2.88), we multiply both sides of the equation by $\sin(\lambda_m\tilde{x})$ and integrate the resulting expressions across the region of interest for $\tilde{x}$ between zero and one. This results in

$$\int_0^1 \left(\Theta_0(\tilde{x}) - \tilde{x}\right)\sin(\lambda_m\tilde{x})d\tilde{x} = \int_0^1 \sum_{n=1}^{\infty} C_n \sin(\lambda_n\tilde{x})\sin(\lambda_m\tilde{x})d\tilde{x} \tag{2.89}$$

Exchanging the summation and integration signs on the right hand side of this equation leads to

$$\int_0^1 \left(\Theta_0(\tilde{x}) - \tilde{x}\right)\sin(\lambda_m\tilde{x})d\tilde{x} = \sum_{n=1}^{\infty} C_n \int_0^1 \sin(\lambda_m\tilde{x})\sin(\lambda_n\tilde{x})d\tilde{x} \tag{2.90}$$

If we now evaluate the integrals on the right side of Eq. (2.90), we find that

$$\int_0^1 \sin(\lambda_m\tilde{x})\sin(\lambda_n\tilde{x})d\tilde{x} \quad = 0 \quad \text{for } n \neq m \tag{2.91}$$

$$= \int_0^1 \sin^2(\lambda_n\tilde{x})d\tilde{x} \quad = \frac{1}{2} \quad \text{for } n = m$$

Writing Eq. (2.90) in detail gives

$$\int_0^1 \left(\Theta_0(\tilde{x}) - \tilde{x}\right)\sin(m\pi\tilde{x})d\tilde{x} \quad = C_1 \int_0^1 \sin(m\pi\tilde{x})\sin(\pi\tilde{x})d\tilde{x} \tag{2.90}$$

$$+C_2 \int_0^1 \sin(m\pi\tilde{x})\sin(2\pi\tilde{x})d\tilde{x}$$

$$+$$

$$+C_n \int_0^1 \sin^2(n\pi\tilde{x})d\tilde{x} \quad (\text{for } n = m)$$

$$+$$

$$+C_\alpha \int_0^1 \sin(m\pi\tilde{x})\sin(\alpha\pi\tilde{x})d\tilde{x}$$

$$+ \dots$$

From Eq. (2.91) one can see that in the sum on the right hand side of this equation only the term containing C_n will be non zero. Therefore, Eq. (2.90) reduces to

$$\int_0^1 (\Theta_0(\tilde{x}) - \tilde{x})\sin(n\pi\tilde{x})d\tilde{x} = C_n \int_0^1 \sin^2(n\pi\tilde{x})d\tilde{x} \tag{2.92}$$

From Eq. (2.92), the unknown constants C_n can be evaluated to be

$$C_n = \frac{\int_0^1 (\Theta_0(\tilde{x}) - \tilde{x})\sin(n\pi\tilde{x})d\tilde{x}}{\int_0^1 \sin^2(n\pi\tilde{x})d\tilde{x}} = 2\int_0^1 (\Theta_0(\tilde{x}) - \tilde{x})\sin(n\pi\tilde{x})d\tilde{x} \tag{2.93}$$

The final solution of the problem is given by Eq. (2.66) with the steady-state solution according to Eq. (2.67) and the transient solution according to Eq. (2.87). Thus,

$$\Theta = \tilde{x} + \sum_{n=1}^{\infty} C_n \sin(n\pi\tilde{x})\exp\left(-(n\pi)^2 \tilde{t}\right) \tag{2.94}$$

In order to show the transient evolution of the temperature field, we select $\Theta_0(\tilde{x}) = 1$ for the above example. Using this initial condition, Eq. (2.93) becomes

$$C_n = 2\int_0^1 (1 - \tilde{x})\sin(n\pi\tilde{x})d\tilde{x} = \frac{2}{n\pi} \tag{2.95}$$

and the temperature distribution in the solid is given by

$$\Theta = \tilde{x} + \sum_{n=1}^{\infty} \frac{2}{n\pi} \sin(n\pi\tilde{x}) \exp\left(-(n\pi)^2 \tilde{t}\right) \qquad (2.96)$$

This temperature distribution is shown in Fig. 2.7 for different times. One can see nicely that the temperature distribution in the solid changes from the prescribed constant initial temperature distribution to the linear temperature shape for the steady-state temperature distribution for $\tilde{t} \to \infty$.

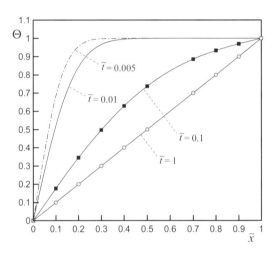

Fig. 2.7: Transient temperature distribution in the plane wall for selected times

In addition, one can see from Eq. (2.96) that the individual parts of the sum in this equation are decaying rapidly with increasing time (notice that the argument of the exponential function contains $(n\pi)^2$ as a multiplier).

2.3.2 Steady-State Heat Conduction in a Rectangular Plate

As a second example to explain the method of separation of variables, we investigate the heat conduction in a rectangular plate, which has height c and width b (see Fig. 2.8).

All four sides of the plate are set to a constant temperature T_0. Inside the rectangular area, a sink is located with constant sink intensity K. Assuming constant physical properties, the energy equation for this steady-state heat conduction problem is given by

$$0 = k\left(\frac{\partial^2 T}{\partial x^2} + \frac{\partial^2 T}{\partial y^2}\right) + K \qquad (2.97)$$

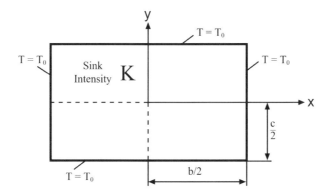

Fig. 2.8: Geometrical configuration and boundary conditions for the heat conduction problem in a flat plate

with boundary conditions

$$T(b/2, y) = T_0, \qquad T(-b/2, y) = T_0 \qquad (2.98)$$
$$T(x, c/2) = T_0, \qquad T(x, -c/2) = T_0$$

As for the first example, before proceeding with the solution of the problem, we first introduce dimensionless quantities. Suitable dimensionless quantities are given by

$$\Theta = \frac{T - T_0}{T_0}, \quad \tilde{x} = \frac{x}{c/2}, \quad \tilde{y} = \frac{y}{c/2} \qquad (2.99)$$

Introducing these quantities into Eqs. (2.97-2.98) results in

$$0 = \frac{\partial^2 \Theta}{\partial \tilde{x}^2} + \frac{\partial^2 \Theta}{\partial \tilde{y}^2} + \overline{K} \qquad (2.100)$$

$$\Theta(A, \tilde{y}) = 0, \qquad \Theta(-A, \tilde{y}) = 0 \qquad (2.101)$$
$$\Theta(\tilde{x}, 1) = 0, \qquad \Theta(\tilde{x}, -1) = 0$$

where the following abbreviations have been used:

$$\overline{K} = \frac{K c^2}{4kT_0}, \quad A = \frac{b}{c} \qquad (2.102)$$

This problem is described by an elliptic second-order partial differential equation. The boundary conditions, expressed by Eq. (2.101), are homogeneous, but the differential equation (2.100) is not.

In order to find a solution of the problem, we make once again use of the method of superposition, since the partial differential equation is linear, and split the solution into two parts

$$\Theta = \Theta_h + \Theta_p \tag{2.103}$$

Θ_h represents the solution of the homogeneous differential equation (without a sink, $K = 0$) and Θ_p is one particular solution of the problem. Let us first focus on the special solution of the problem. In order to find this solution, we assume that $\Theta_p = f(\tilde{y})$. Alternatively, we could also assume $\Theta_p = f(\tilde{x})$ and obtain the same final solution of the problem. The analysis, however, would be altered.

If we substitute $\Theta_p = f(\tilde{y})$, into Eq. (2.100), the following relation for the function $f(\tilde{y})$ is obtained

$$f''(\tilde{y}) = -\bar{K} \quad \Rightarrow f(\tilde{y}) = -\frac{1}{2}\bar{K}\,\tilde{y}^2 + C_1\tilde{y} + C_2 \tag{2.104}$$

Since we need only one special solution of the problem, we could set C_1 and C_2 equal to zero. However, a better choice is to select the constants C_1 and C_2 in such a way that the two boundary conditions $\Theta_p(\tilde{x},1) = 0, \quad \Theta_p(\tilde{x},-1) = 0$ are satisfied. If we do so, we obtain the following solution of the problem

$$\Theta_p = \frac{\bar{K}}{2}\left[1 - \tilde{y}^2\right] \tag{2.105}$$

After having obtained the solution for Θ_p, one has to solve the following problem for Θ_h which is deduced from the Eqs. (2.100 –2.101)

$$0 = \frac{\partial^2 \Theta_h}{\partial \tilde{x}^2} + \frac{\partial^2 \Theta_h}{\partial \tilde{y}^2} \tag{2.106}$$

$$\Theta_h(A,\tilde{y}) = -\bar{K}/2\left(1 - \tilde{y}^2\right), \qquad \Theta_h(-A,\tilde{y}) = -\bar{K}/2\left(1 - \tilde{y}^2\right) \tag{2.107}$$
$$\Theta_h(\tilde{x},1) = 0, \qquad\qquad \Theta_h(\tilde{x},-1) = 0$$

Note that the differential equation for Θ_h is now homogeneous, whereas two of the boundary conditions are non-homogeneous. Furthermore, note that the two boundary conditions, corresponding to a constant value of $\tilde{y}$, are still homogeneous. This is of importance for the subsequent analysis of the problem.

We now assume that Eq. (2.106) has a solution, which can be obtained by the method of separation of variables. Thus

$$\Theta_h = F(\tilde{x})G(\tilde{y}) \tag{2.108}$$

Introducing Eq. (2.108) into the Eqs. (2.106-2.107) results in

$$F''(\tilde{x})\ G(\tilde{y}) + G''(\tilde{y})\ F(\tilde{x}) = 0 \qquad\qquad (2.109)$$

from which we obtain

$$\frac{F''(\tilde{x})}{F(\tilde{x})} = -\frac{G''(\tilde{y})}{G(\tilde{y})} = \pm\lambda^2 \qquad\qquad (2.110)$$

From Eq. (2.110), one notices that a physically plausible solution occurs for both $+\lambda^2$ and $-\lambda^2$. In order to investigate this problem further, we analyse in more details the solutions for the function $G(\tilde{y})$. From Eq. (2.110) we get

$$G_1''(\tilde{y}) + \lambda^2\, G_1(\tilde{y}) = 0 \quad \text{for } +\lambda^2 \qquad\qquad (2.111)$$

$$G_2''(\tilde{y}) - \lambda^2 G_2(\tilde{y}) = 0 \quad \text{for } -\lambda^2 \qquad\qquad (2.112)$$

which gives rise to the following two possible solutions

$$G_1(\tilde{y}) = C_3 \cos(\lambda\tilde{y}) + C_4 \sin(\lambda\tilde{y}) \qquad\qquad (2.113)$$

$$G_2(\tilde{y}) = C_3 \cosh(\lambda\tilde{y}) + C_4 \sinh(\lambda\tilde{y}) \qquad\qquad (2.114)$$

If we now reconsider the problem to be solved (Eqs. (2.106-2.107)), one can see that $\Theta_h = F(\tilde{x})\,G(\tilde{y})$ has to be zero for $\tilde{y} = \pm 1$. If we satisfy these two boundary conditions by Eq. (2.114), we obtain only the trivial solution, because the functions $\cosh(\lambda\tilde{y})$ and $\sinh(\lambda\tilde{y})$ have only one zero point. Instead, if we satisfy the two boundary conditions by Eq. (2.113), we obtain an equation, which determines the eigenvalues. Therefore, Eq. (2.113) is the desired solution and thus, we have to select $+\lambda^2$ in Eq. (2.110).

For the function $F(\tilde{x})$, one obtains from Eq. (2.110)

$$F''(\tilde{x}) - \lambda^2 F(\tilde{x}) = 0 \qquad\qquad (2.115)$$

which has the solution

$$F(\tilde{x}) = C_5 \cosh(\lambda\tilde{x}) + C_6 \sinh(\lambda\tilde{x}) \qquad\qquad (2.116)$$

Combining the solutions for F and G leads to the following expression for Θ_h

$$\Theta_h = \left(C_3 \cos(\lambda\tilde{y}) + C_4 \sin(\lambda\tilde{y})\right)\left(C_5 \cosh(\lambda\tilde{x}) + C_6 \sinh(\lambda\tilde{x})\right) \qquad (2.117)$$

This expression has to satisfy the boundary conditions given by Eq. (2.107)

$$\Theta_h\left(A,\tilde{y}\right) = -\overline{K}/2\left(1 - \tilde{y}^2\right), \qquad \Theta_h\left(-A,\tilde{y}\right) = -\overline{K}/2\left(1 - \tilde{y}^2\right) \qquad (2.107)$$

$$\Theta_h\left(\tilde{x},1\right) = 0, \qquad\qquad\qquad \Theta_h\left(\tilde{x},-1\right) = 0$$

Applying the two boundary conditions for fixed values of $\tilde{y}$, the following two equations are obtained

$$C_3 \cos(\lambda) \qquad + C_4 \sin(\lambda) \qquad = 0$$
$$C_3 \cos(\lambda(-1)) + C_4 \sin(\lambda(-1)) = 0$$

(2.118)

Because $\cos(\lambda) = \cos(-\lambda)$ and $\sin(\lambda) = -\sin(-\lambda)$, one obtains from the above equations that $C_4 = 0$ and that

$$C_3 \cos(\lambda) = 0$$

(2.119)

From this equation it follows that

$$\lambda = \frac{2n-1}{2}\pi, \qquad n = 1,2,3,...$$

(2.120)

and the following solution for Θ_h is obtained

$$\Theta_h = C_3 \cos(\lambda\tilde{y}) \ (C_5 \cosh(\lambda\tilde{x}) + C_6 \sinh(\lambda\tilde{x}))$$

(2.121)

From the two boundary conditions, corresponding to a fixed value of $\tilde{x}$ in Eq. (2.107), it can be seen that $\Theta_h(A, \tilde{y}) = \Theta_h(-A, \tilde{y})$, which indicates Θ_h is an even function in $\tilde{x}$. Therefore, it follows that $C_6 = 0$. Thus

$$\Theta_h = C \cos(\lambda\tilde{y})\cosh(\lambda\tilde{x}) , \qquad C = C_3 C_5$$

(2.122)

Since an infinite number of eigenvalues has been found from Eq. (2.120), the solution for Θ_h can be constructed by superimposing all these individual solutions.

This results in

$$\Theta_h = \sum_{n=1}^{\infty} C_n \cos(\lambda_n\tilde{y})\cosh(\lambda_n\tilde{x})$$

(2.123)

The unknown coefficients C_n can be obtained by matching the boundary condition $\Theta_h(A, \tilde{y}) = -\overline{K}/2(1-\tilde{y}^2)$ by Eq. (2.122). This results in

$$-\frac{\overline{K}}{2}(1-\tilde{y}^2) = \sum_{n=1}^{\infty} C_n \cos(\lambda_n\tilde{y})\cosh(\lambda_n A)$$

(2.124)

If we multiply both sides of the above equation by $\cos(\lambda_m\tilde{y})$ and integrate the resulting expressions between -1 and 1 we obtain:

$$-\frac{\overline{K}}{2}\int_{-1}^{1}(1-\tilde{y}^2)\cos(\lambda_m\tilde{y})d\tilde{y} = \int_{-1}^{1}\left(\sum_{n=1}^{\infty} C_n \cos(\lambda_n\tilde{y})\cos(\lambda_m\tilde{y})\cosh(\lambda_n A)\right)d\tilde{y}$$

(2.125)

Again, the summation and integration signs on the right hand side of Eq. (2.125) can be interchanged. Then, it is obvious that from the sum only one term will not be equal to zero, because the integral

$$\int_{-1}^{1} \cos(\lambda_n \tilde{y}) \cos(\lambda_m \tilde{y}) d\tilde{y} \quad = \quad 0 \quad \text{for } \lambda_n \neq \lambda_m \tag{2.126}$$

$$= \quad 1 \quad \text{for } \lambda_n = \lambda_m$$

Finally, the following equation is obtained for the determination of the unknown coefficients

$$C_n = \frac{-\dfrac{\overline{K}}{2} \displaystyle\int_{-1}^{1} (1 - \tilde{y}^2) \cos(\lambda_n \tilde{y}) d\tilde{y}}{\cosh(\lambda_n A)} = \frac{2\overline{K}(-1)^n}{\lambda_n^3 \cosh(\lambda_n A)} \tag{2.127}$$

The solution of the problem, given by Eqs. (2.100)-(2.101), is obtained by combining the two parts of the solution Θ_h and Θ_p. This gives finally

$$\Theta = \frac{\overline{K}}{2}\left(1 - \tilde{y}^2 + \sum_{n=1}^{\infty} \frac{4(-1)^n}{\lambda_n^3 \cosh(\lambda_n A)} \cos(\lambda_n \tilde{y}) \cosh(\lambda_n \tilde{x})\right) \tag{2.128}$$

The solution obtained here shall serve as an example that a lot of solutions, which are obtained for heat conduction problems, can be very useful for other applications. To that aim, let us investigate the flow in a rectangular channel. The channel has width b and height c. The geometry under consideration is shown in Fig. 2.9.

For this problem, u, v, w are the flow velocities in the x,y and z direction. Under the assumption of a steady, incompressible, laminar flow with constant fluid properties, the Navier-Stokes equations reduce to

$$\rho\left(u\frac{\partial u}{\partial x} + v\frac{\partial u}{\partial y} + w\frac{\partial u}{\partial z}\right) = F_x - \frac{\partial p}{\partial x} + \mu\left(\frac{\partial^2 u}{\partial x^2} + \frac{\partial^2 u}{\partial y^2} + \frac{\partial^2 u}{\partial z^2}\right) \tag{1.1}$$

$$\rho\left(u\frac{\partial v}{\partial x} + v\frac{\partial v}{\partial y} + w\frac{\partial v}{\partial z}\right) = F_y - \frac{\partial p}{\partial y} + \mu\left(\frac{\partial^2 v}{\partial x^2} + \frac{\partial^2 v}{\partial y^2} + \frac{\partial^2 v}{\partial z^2}\right) \tag{1.2}$$

$$\rho\left(u\frac{\partial w}{\partial x} + v\frac{\partial w}{\partial y} + w\frac{\partial w}{\partial z}\right) = F_z - \frac{\partial p}{\partial z} + \mu\left(\frac{\partial^2 w}{\partial x^2} + \frac{\partial^2 w}{\partial y^2} + \frac{\partial^2 w}{\partial z^2}\right) \tag{1.3}$$

and the mass continuity equation is

$$\frac{\partial u}{\partial x} + \frac{\partial v}{\partial y} + \frac{\partial w}{\partial z} = 0 \tag{1.6}$$

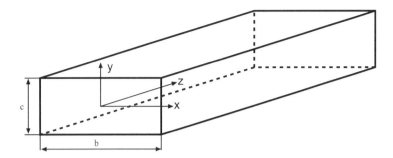

Fig. 2.9: Geometrical configuration and coordinate system for the flow in a rectangular channel

If we now further assume that the flow is hydrodynamically fully developed, i.e. the velocity profile does not change along the axial direction, only the w component of the flow velocity is present. The u and v components are identically zero. Additionally, the w component of the flow can only be a function of the x and y coordinate for the hydrodynamically fully developed flow. Then the problem simplifies to (see for example Spurk (1987))

$$0 = -\frac{\partial p}{\partial z} + \mu \left(\frac{\partial^2 w}{\partial x^2} + \frac{\partial^2 w}{\partial y^2} \right) \tag{2.129}$$

where the pressure gradient in the axial direction is constant for a hydrodynamically fully developed flow. The boundary conditions are the no slip conditions at all boundaries of the channel. Thus

$$w(b/2, y) = 0, \qquad w(-b/2, y) = 0 \tag{2.130}$$
$$w(x, c/2) = 0, \qquad w(x, -c/2) = 0$$

If we introduce into Eq. (2.129) the abbreviation $\bar{K} = -1/\mu \, \partial p / \partial z$, it can be seen that the problem for determining the fully developed velocity profile is identical to the heat conduction problem in a plate containing a heat sink with constant sink intensity (the derivation of the fully developed velocity field in a rectangular channel is given for example in Spurk (1987)). This shows nicely the similarity of the equations describing problems in Fluid Mechanics and Heat Transfer.

2.3.3 Separation of Variables for the General Case of a Linear Second-Order Partial Differential Equation

At the beginning of this chapter, we have been concerned with a linear second order partial differential equation which depended on the two variables x and y. The most general form of this homogeneous equation is given by

$$A(x,y)\frac{\partial^2 u}{\partial x^2}+2B(x,y)\frac{\partial^2 u}{\partial x\partial y}+C(x,y)\frac{\partial^2 u}{\partial y^2}+ \tag{2.1}$$

$$+D(x,y)\frac{\partial u}{\partial x}+E(x,y)\frac{\partial u}{\partial y}+F(x,y)u=0$$

where $A,...,F$ are constants or functions of x and y which are sufficiently differentiable in the domain of interest.

In the previous two examples, we used the method of separation of variables to derive a solution of the linear partial differential equation as an infinite sum. However, we only addressed very special cases of Eq. (2.1). It is now interesting to evaluate, under which conditions a separation of variables is possible for Eq. (2.1). In order to answer this question, we consider the transformed equation (2.16)

$$\bar{A}(\xi,\eta)\frac{\partial^2 u}{\partial \xi^2}+2\bar{B}(\xi,\eta)\frac{\partial^2 u}{\partial \xi\partial \eta}+\bar{C}(\xi,\eta)\frac{\partial^2 u}{\partial \eta^2}+ \tag{2.16}$$

$$\bar{D}(\xi,\eta)\frac{\partial u}{\partial \xi}+\bar{E}(\xi,\eta)\frac{\partial u}{\partial \eta}+F(\xi,\eta)u=0$$

where the new coordinates ξ, η, defined by Eq. (2.6), have been used. Let us substitute

$$u=H(\xi)G(\eta) \tag{2.131}$$

into Eq. (2.16). From this we obtain

$$\bar{A}(\xi,\eta)H''(\xi)G(\eta)+2\bar{B}(\xi,\eta)H'(\xi)G'(\eta)+\bar{C}(\xi,\eta)H(\xi)G''(\eta)+ \tag{2.132}$$
$$\bar{D}(\xi,\eta)H'(\xi)G(\eta)+\bar{E}(\xi,\eta)H(\xi)G'(\eta)+F H(\xi)G(\eta)=0$$

where the prime indicates the differentiation of the functions $H(\xi)$ and $G(\eta)$ with respect to the independent variable. Dividing Eq. (2.132) by $H(\xi)G(\eta)$ results in

$$\bar{A}(\xi,\eta)\frac{H''(\xi)}{H(\xi)}+2\bar{B}(\xi,\eta)\frac{H'(\xi)}{H(\xi)}\frac{G'(\eta)}{G(\eta)}+\bar{C}(\xi,\eta)\frac{G''(\eta)}{G(\eta)}+ \tag{2.133}$$

$$\bar{D}(\xi,\eta)\frac{H'(\xi)}{H(\xi)}+\bar{E}(\xi,\eta)\frac{G'(\eta)}{G(\eta)}+F(\xi,\eta)=0$$

From the above equation it can be seen that the variables can only be separated if $\bar{B}(\xi,\eta)=0$. According to Eq. (2.17), this requires that the new coordinates are chosen in a way to ensure that

$$\bar{B}=A\frac{\partial \xi}{\partial x}\frac{\partial \eta}{\partial x}+B\left(\frac{\partial \xi}{\partial x}\frac{\partial \eta}{\partial y}+\frac{\partial \xi}{\partial y}\frac{\partial \eta}{\partial x}\right)+C\frac{\partial \xi}{\partial y}\frac{\partial \eta}{\partial y}=0 \tag{2.134}$$

After setting $\bar{B}(\xi,\eta) = 0$, Eq. (2.132) can be rewritten in the following way

$$\frac{\bar{A}(\xi,\eta)}{N(\xi,\eta)}\frac{H''(\xi)}{H(\xi)} + \frac{\bar{D}(\xi,\eta)}{N(\xi,\eta)}\frac{H'(\xi)}{H(\xi)} + \frac{\bar{C}(\xi,\eta)}{N(\xi,\eta)}\frac{G''(\eta)}{G(\eta)} + \tag{2.135}$$

$$+\frac{\bar{E}(\xi,\eta)}{N(\xi,\eta)}\frac{G'(\eta)}{G(\eta)} + \frac{F(\xi,\eta)}{N(\xi,\eta)} = 0$$

where the whole equation has been divided by the function $N(\xi,\eta)$. If we further assume that

$$\frac{F(\xi,\eta)}{N(\xi,\eta)} = f_1(\xi) + f_2(\eta) \tag{2.136}$$

Eq. (2.134) can be written as

$$\frac{\bar{A}(\xi,\eta)}{N(\xi,\eta)}\frac{H''(\xi)}{H(\xi)} + \frac{\bar{D}(\xi,\eta)}{N(\xi,\eta)}\frac{H'(\xi)}{H(\xi)} + f_1(\xi) = \tag{2.137}$$

$$-\frac{\bar{C}(\xi,\eta)}{N(\xi,\eta)}\frac{G''(\eta)}{G(\eta)} - \frac{\bar{E}(\xi,\eta)}{N(\xi,\eta)}\frac{G'(\eta)}{G(\eta)} - f_2(\eta) = \text{const.}$$

The left side of this equation should now only be a function of ξ and the right side of the equation should only be a function of η. This is only possible, if the following restrictions are satisfied:

$$\frac{\bar{A}(\xi,\eta)}{N(\xi,\eta)} = f_3(\xi), \quad \frac{\bar{D}(\xi,\eta)}{N(\xi,\eta)} = f_4(\xi), \tag{2.138}$$

$$\frac{\bar{C}(\xi,\eta)}{N(\xi,\eta)} = f_5(\eta), \quad \frac{\bar{E}(\xi,\eta)}{N(\xi,\eta)} = f_6(\eta)$$

The present analysis might be very helpful in order to check in advance if the method of separation of variables can be applied to the problem under consideration. It would be incorrect, however, to assume that the method leads to a solution in all cases, where the separation of variables is possible.

Problems

2.1   Consider the partial differential equation

$$4\frac{\partial^2 u}{\partial x^2} + 5\frac{\partial^2 u}{\partial x \partial y} + \frac{\partial^2 u}{\partial y^2} + \frac{\partial u}{\partial x} + \frac{\partial u}{\partial y} = 2$$

a.)   Determine the type of the differential equation.

b.) What are the characteristics of this equation?

c.) Transform the equation into its standard form.

d.) Determine the general solution of the above given differential equation (hint: use the substitution: $v = \partial u / \partial \eta$).

2.2   Consider the partial differential equation

$$-x\frac{\partial^2 u}{\partial x^2}+y\frac{\partial^2 u}{\partial y^2}-\frac{1}{2}\frac{\partial u}{\partial x}+A(x,y)\frac{\partial u}{\partial y}=0$$

Solve this equation using the method of separation of variables.

a.) Insert $u = F(x)G(x)$ into the equation. How should the function $A(x, y)$ look like, so that the method of separation of variables can lead to a solution of the above equation?

b.) Determine the type of the partial differential equation. Show the result in a x-y diagram for $-\infty < x < +\infty$ and $-\infty < y < +\infty$.

c.) Calculate for $x > 0$, $y > 0$ the characteristics of the equation and transform the equation into its standard form (use for this $A(x, y) = -1/2$).

2-1.   Consider the partial differential equation

$$\frac{\partial^2 u}{\partial x^2}+5\frac{\partial^2 u}{\partial x\partial y}+4\frac{\partial^2 u}{\partial y^2}+10\frac{\partial u}{\partial y}=\sin x$$

a.) Determine the type of the differential equation.

b.) What are the characteristics of this equation?

c.) Transform the equation into its standard form.

2-2.   A thin rectangular plate with sides of length a and b is subjected to a constant temperature T_1 at three sides $(T(x,0)=T(x,b)=T(0,y)=T_1)$, whereas the remaining side of this plate is subjected to the temperature distribution

$$T(a, y) = T_1\left\{\sin^3\left(\frac{2\pi y}{b}\right)+1\right\}$$

We are interested in the steady-state temperature distribution in the plate. All material properties of the plate are constant. The steady-state temperature distribution can be calculated from the energy equation

$$\frac{\partial^2 T}{\partial x^2}+\frac{\partial^2 T}{\partial y^2}=0$$

with the above given boundary conditions.

a.) Make the energy equation and the boundary conditions dimensionless.
b.) Solve the problem using the method of separation of variables.

2-3. Consider a slab, which has extensions in the y- and z-direction much bigger than in the x- direction. The slab has constant initial temperature T_i.
At $t = 0$, the slab is exposed at both sides ($x = 0, x = \delta$) to convective
cooling. The temperatures of the surrounding fluid are given by T_C and
T_G. The problem under consideration is described by the following, simplified energy equation

$$\rho c \frac{\partial T}{\partial t} = k \frac{\partial^2 T}{\partial x^2}$$

and the following boundary conditions

$$t = 0 : T = T_i$$

$$x = 0 : h_G (T_G - T(x = 0)) + k \frac{\partial T}{\partial x} \bigg|_{x=0} = 0$$

$$x = \delta : h_G (T_C - T(x = \delta)) - k \frac{\partial T}{\partial x} \bigg|_{x=\delta} = 0$$

a.) Introduce dimensionless quantities into the above equations. Use

$$\Theta = \frac{T_G - T}{T_G - T_i}, \ \tilde{t} = \frac{at}{\delta^2}, \ a = \frac{k}{\rho c}, \ \tilde{x} = \frac{x}{\delta},$$

$$Bi_G = \frac{h_G \delta}{k}, \ Bi_C = \frac{h_C \delta}{k}$$

b.) Split the solution of the problem into the steady-state solution and
into the solution of the transient part. Show that for the transient part
of the solution, the two boundary conditions for $\tilde{x} = 0$ and $\tilde{x} = 1$ are
homogeneous.
c.) Solve the first problem. What is the steady-state temperature distribution in the slab?
d.) Solve the transient problem. What is the complete solution of the
problem?

2-4. Consider the transient heat conduction in a slab of length l. The slab has
the initial temperature distribution

$$T(0, x) = \frac{x - l}{l} (T_2 - T_1) + T_1$$

At both sides of the slab, the following constant temperatures are applied

$$T(t,0) = T_1, \quad T(t,l) = T_2$$

In addition, the slab contains a heat source. The above given problem can be described by the following partial differential equation (where a and B are constants)

$$\frac{\partial T}{\partial t} = a \frac{\partial^2 T}{\partial x^2} + \frac{xBa(T_2 - T_1)}{l^3}$$

a.) Make the differential equation and the boundary conditions dimensionless by introducing suitable variables.
b.) Split the problem into different simpler problems.
c.) Solve the different problems and give the complete solution.

2-5.   Consider a sphere with radius R. For $t = 0$ the sphere has constant temperature T_i. The surface of the sphere is set to the constant temperature T_0 for $t > 0$. The sphere contains also a heat source with constant source intensity $\dot{q}_{i0}$. The material properties of the sphere are considered to be constant. The temperature distribution in the sphere can be calculated from the energy equation in spherical coordinates

$$\rho c \frac{\partial T}{\partial t} = \frac{k}{r^2} \frac{\partial}{\partial r}\left(r^2 \frac{\partial T}{\partial r}\right) + \frac{k}{r^2 \sin \psi} \frac{\partial}{\partial \psi}\left(\sin \psi \frac{\partial T}{\partial \psi}\right) +$$

$$+ \frac{k}{r^2 \sin \psi} \frac{\partial}{\partial \varphi}\left(\frac{\partial T}{\partial \varphi}\right) + \dot{q}_{i0}$$

a.) Simplify the above given energy equation for the case of rotational symmetry. What are the needed boundary conditions?
b.) Transform the problem under consideration by using $T(r,t) = U(r,t)/r$. What is the resulting differential equation and what are the boundary conditions?
c.) Introduce dimensionless quantities, so that the spatial boundary conditions are homogeneous.
d.) Solve the transformed problem by using the method of separation of variables.
e.) Derive from the solution given under d.) the temperature distribution $T(r,t)$.

3 Heat Transfer in Pipe and Channel Flows (Parabolic Problems)

The method of separation of variables has been briefly explained in the last chapter by solving two simple examples. In general, this solution method leads to eigenvalue problems which have to be solved either analytically or numerically. For most technical problems, where analytical solutions based on the method of separation of variables are still obtainable, these eigenvalue problems might become quite difficult.

In order to show how the method can be applied to more complicated problems, the present chapter focuses on the solution of heat transfer problems in pipe and channel flows with hydrodynamically fully developed velocity profiles. In Chap. 3, we will also explore some of the general properties of the eigenvalue problems under consideration.

3.1 Heat Transfer in Pipe and Channel Flows with Constant Wall Temperature

As a first example, we investigate the technical important and scientific very interesting problem of convective heat transfer in a turbulent flow of a liquid, flowing in a pipe or a planar channel. We assume that the flow is hydrodynamically fully developed and that the fluid properties are constant. Due to their technical and scientific relevance, these types of flows have been investigated in great detail in the past. Good reviews can be found in Tietjens (1970), Schlichting (1982) and in Bhatti and Shah (1987).

Heat transfer problems in laminar pipe or channel flows are also of great theoretical and practical interest. Due to the simplicity of the velocity profile for the hydrodynamically fully developed flow, the heat transfer characteristics for the thermal entry length have been investigated relatively early by Graetz (1883, 1885) and independently by Nusselt (1910). This is the reason that this kind of problems are sometimes referred in literature as Graetz problems. For the heat transfer characteristics in laminar pipe and channel flows, a large number of publications does exist. Good reviews can be found in Shah and London (1978) and in Bhatti and Shaw (1987). In particular, with respect to the thermal development of a hydrodynamically fully developed laminar flow, several theoretical and numerical methods have been evaluated and compared. The reader is referred to Shah and London (1978) for more details. On the other hand, the thermal entrance of a hydrodynamically fully developed turbulent pipe or channel flow has a much broader technical relevance and has not yet been investigated in such detail in the past. Furthermore, this kind of application leads to complicated eigenvalue problems and can be considered as a challenging case for any analytical method. Of course, the present analysis holds also for laminar pipe and channel flows, as stressed in the following sections when relevant. In addition, the program for the solution of the eigenvalue problems, described in Appendix C, can be used for both laminar and turbulent flows.

3.1.1 Velocity Distribution of Hydrodynamically Fully Developed Pipe and Channel Flows

Figure 3.1 shows the geometry and the coordinate system of the problem under consideration. We assume a hydrodynamically fully developed flow, which means that the velocity distribution does not change with increasing values of x. This has the implication that only the velocity component u in the x-direction is nonzero. Furthermore, u is only a function of the coordinate orthogonal to the flow direction. This coordinate is y in case of a planar channel and r in case of a circular pipe. In the following the pipe radius is denoted by R and the distance between the two parallel plates by $2h$.

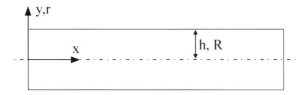

Fig. 3.1: Geometry and coordinate system

Furthermore, we assume that the fluid properties are constant. This is a suitable assumption if the temperature differences in the problem are not to large. Then the velocity distribution for the pipe and channel flow can be calculated analytically from the momentum equations. A detailed description is given in Appendix A. For a laminar flow, very simple expressions are obtained:

Laminar Pipe Flow

$$\frac{u}{\bar{u}} = 2\left(1 - \left(\frac{r}{R}\right)^2\right) \tag{3.1}$$

Laminar Flow in a Planar Channel

$$\frac{u}{\bar{u}} = \frac{3}{2}\left(1 - \left(\frac{y}{h}\right)^2\right)$$
(3.2)

where $\bar{u}$ is the mean velocity of the pipe or channel flow. This mean velocity can be calculated from the measured flow rate in the pipe or channel.

For a turbulent hydrodynamically fully developed pipe or channel flow, the turbulent shear stress in the momentum equation in the x-direction has to be modelled by using a turbulence model. For pipe and channel flows, this can be done efficiently by using a simple mixing length model (Cebeci and Bradshaw (1984), Cebeci and Chang (1978), Schlichting (1982)). After some algebra, one obtains finally a description of the velocity distribution in the pipe or in the channel of the following form (the reader is referred to Appendix A for a detailed derivation of the Eqs. (3.3-3.4)).

Turbulent Pipe Flow

$$\frac{u}{\bar{u}} = f\left(\frac{r}{R}, \mathrm{Re}_R\right)$$
(3.3)

Turbulent Flow in a Planar Channel

$$\frac{u}{\bar{u}} = f\left(\frac{y}{h}, \mathrm{Re}_h\right)$$
(3.4)

Obviously, in a turbulent flow the velocity distribution depends on the Reynolds number. Increasing values of the Reynolds number lead to an enhancement of the turbulent mixing within the cross sectional area of the duct and, therefore, to a flatter velocity profile.

Figure 3.2 shows predicted and measured velocity profiles for hydrodynamically fully developed turbulent flows in a pipe and in a planar channel, respectively. In Fig. 3.2, the velocity is scaled by its maximum value in the center. The figure on the left hand side shows a calculation for the flow in a planar channel. It can be seen that the predictions using the simple mixing length model agree quite well with the measurements of Laufer (1950). The hydrodynamically fully developed velocity profile in a pipe is depicted on the right hand side of Fig. 3.2. It can be seen that the velocity profile changes its shape by increasing the Reynolds number from 2.3 10^4 to 3.2 10^6. In addition, the good agreement between calculations and measurements of Nikuradse (1932) can be observed (see also Appendix A).

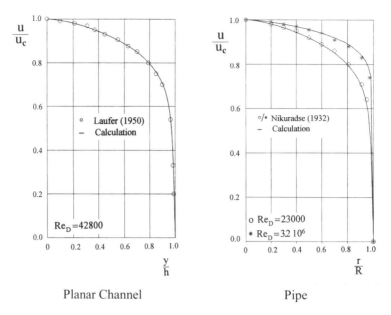

Planar Channel   Pipe

Fig. 3.2: Hydrodynamically fully developed velocity distribution in a pipe and in a planar channel for different Reynolds numbers

3.1.2 Thermal Entrance Solutions for Constant Wall Temperature

Once the velocity profile is known, the energy equation can be analyzed. If we assume an incompressible flow with constant fluid properties and neglect viscous dissipation, the energy equation takes the following form for the hydrodynamically fully developed turbulent flow:

Pipe

$$\rho c_p u(r)\frac{\partial T}{\partial x} = \frac{1}{r}\frac{\partial}{\partial r}\left[r\left(k\frac{\partial T}{\partial r} - \rho c_p \overline{v'T'} \right) \right] + \frac{\partial}{\partial x}\left[k\frac{\partial T}{\partial x} - \rho c_p \overline{u'T'} \right] \qquad (3.5)$$

Planar Channel

$$\rho c_p u(y)\frac{\partial T}{\partial x} = \frac{\partial}{\partial y}\left[k\frac{\partial T}{\partial y} - \rho c_p \overline{v'T'} \right] + \frac{\partial}{\partial x}\left[k\frac{\partial T}{\partial x} - \rho c_p \overline{u'T'} \right] \qquad (3.6)$$

Figure 3.3 shows the geometry and the boundary conditions. It can be seen that at $x = 0$ the wall temperature is suddenly increased from T_0 to T_W. The fluid temperature has a uniform value T_0 for $x \rightarrow -\infty$. Far away from the entrance ($x \rightarrow \infty$), the fluid temperature will attain asymptotically the uniform wall temperature T_W.

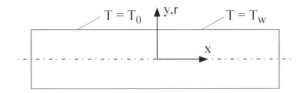

Fig. 3.3: Geometry and boundary conditions

For laminar flow ($\overline{u'T'} = 0$, $\overline{v'T'} = 0$),the equations (3.5) or (3.6) are elliptic in nature. This is caused by the second term on the right hand side of these equations, representing the axial heat conduction effect within the flow. For turbulent flow, the nature of the equation depends also on the turbulent heat fluxes. The turbulent heat fluxes $-\rho c_p \overline{v'T'}$ and $-\rho c_p \overline{u'T'}$ have to be modelled. This can be done for example by using a simple eddy viscosity model:

$$-\overline{v'T'} = \varepsilon_{hy} \frac{\partial T}{\partial y} \text{ (planar channel)}, \qquad -\overline{v'T'} = \varepsilon_{hr} \frac{\partial T}{\partial r} \text{ (pipe)} \qquad (3.7)$$

$$-\overline{u'T'} = \varepsilon_{hx} \frac{\partial T}{\partial x} \text{ (planar channel and pipe)} \qquad (3.8)$$

where ε_{hy}, ε_{hr} and ε_{hx} are only functions of the coordinate orthogonal to the flow direction. Inserting Eqs. (3.7-3.8) into Eq. (3.5) and Eq. (3.6) results in

Pipe

$$\rho c_p u(r) \frac{\partial T}{\partial x} = \frac{1}{r} \frac{\partial}{\partial r}\left[r\left(k + \rho c_p \varepsilon_{hr}\right) \frac{\partial T}{\partial r}\right] + \frac{\partial}{\partial x}\left[\left(k + \rho c_p \varepsilon_{hx}\right) \frac{\partial T}{\partial x}\right] \qquad (3.9)$$

Planar Channel

$$\rho c_p u(y) \frac{\partial T}{\partial x} = \frac{\partial}{\partial y}\left[\left(k + \rho c_p \varepsilon_{hy}\right) \frac{\partial T}{\partial y}\right] + \frac{\partial}{\partial x}\left[\left(k + \rho c_p \varepsilon_{hx}\right) \frac{\partial T}{\partial x}\right] \qquad (3.10)$$

Introducing the following dimensionless quantities into the above equations,

$$\Theta = \frac{T - T_W}{T_0 - T_W}, \quad \tilde{x} = \frac{x}{L} \frac{1}{\mathrm{Re}_L \, \mathrm{Pr}}, \quad \tilde{y} = \frac{y}{L}, \quad \tilde{r} = \frac{r}{L}, \quad \mathrm{Re}_L = \frac{\bar{u} L}{\nu}, \qquad (3.11)$$

$$\tilde{\varepsilon}_{hx} = \frac{\varepsilon_{hx}}{\nu} , \ \tilde{\varepsilon}_{hr} = \frac{\varepsilon_{hr}}{\nu} , \ \tilde{\varepsilon}_{hy} = \frac{\varepsilon_{hy}}{\nu} , \ \tilde{u} = \frac{u}{\bar{u}} , \ \mathrm{Pe}_L = \mathrm{Re}_L \, \mathrm{Pr} , \ \mathrm{Pr} = \frac{\nu}{a}$$

where the length scale is $L = R$ for pipe flows and $L = h$ for the flow in a planar channel. We obtain

Pipe

$$\tilde{u}(\tilde{r}) \frac{\partial \Theta}{\partial \tilde{x}} = \frac{1}{\tilde{r}} \frac{\partial}{\partial \tilde{r}} \left[\tilde{r} \left(1 + \mathrm{Pr}\, \tilde{\varepsilon}_{hr} \right) \frac{\partial \Theta}{\partial \tilde{r}} \right] + \frac{1}{\mathrm{Pe}_L^2} \frac{\partial}{\partial \tilde{x}} \left[\left(1 + \mathrm{Pr}\, \tilde{\varepsilon}_{hx} \right) \frac{\partial \Theta}{\partial \tilde{x}} \right] \qquad (3.12)$$

Planar Channel

$$\tilde{u}(\tilde{y}) \frac{\partial \Theta}{\partial \tilde{x}} = \frac{\partial}{\partial \tilde{y}} \left[\left(1 + \mathrm{Pr}\, \tilde{\varepsilon}_{hy} \right) \frac{\partial \Theta}{\partial \tilde{y}} \right] + \frac{1}{\mathrm{Pe}_L^2} \frac{\partial}{\partial \tilde{x}} \left[\left(1 + \mathrm{Pr}\, \tilde{\varepsilon}_{hx} \right) \frac{\partial \Theta}{\partial \tilde{x}} \right] \qquad (3.13)$$

For Peclet numbers $\mathrm{Pe}_L > 100$ and a semi-infinite heating length, axial heat conduction in the fluid can be neglected with good accuracy. This means that the second term on the right hand side of the Eqs. (3.12-3.13) can be ignored. This leads to:

Pipe

$$\tilde{u}(\tilde{r}) \frac{\partial \Theta}{\partial \tilde{x}} = \frac{1}{\tilde{r}} \frac{\partial}{\partial \tilde{r}} \left[\tilde{r} \left(1 + \mathrm{Pr}\, \tilde{\varepsilon}_{hr} \right) \frac{\partial \Theta}{\partial \tilde{r}} \right] \qquad (3.14)$$

Planar Channel

$$\tilde{u}(\tilde{y}) \frac{\partial \Theta}{\partial \tilde{x}} = \frac{\partial}{\partial \tilde{y}} \left[\left(1 + \mathrm{Pr}\, \tilde{\varepsilon}_{hy} \right) \frac{\partial \Theta}{\partial \tilde{y}} \right] \qquad (3.15)$$

These two equations can be combined into one relation, if we introduce a vertical coordinate n, which is equal to y for a planar channel and equal to r for pipe flows. This results in

$$\tilde{u}(\tilde{n}) \frac{\partial \Theta}{\partial \tilde{x}} = \frac{1}{\tilde{r}^F} \frac{\partial}{\partial \tilde{n}} \left[\tilde{r}^F \left(1 + \mathrm{Pr}\, \tilde{\varepsilon}_{hn} \right) \frac{\partial \Theta}{\partial \tilde{n}} \right] \qquad (3.16)$$

where the superscript F specifies the geometry and has to be set to 0 for a planar channel and 1 for pipe flows. If we examine the nature of Eq. (3.16), we see that the equation is parabolic. The boundary conditions for Eq. (3.16) are therefore given by

$$\tilde{x} = 0 : \Theta(0,\tilde{n}) = 1$$

$$\tilde{n} = 0 : \left.\frac{\partial \Theta}{\partial \tilde{n}}\right|_{\tilde{n}=0} = 0 \qquad (3.17)$$

$$\tilde{n} = 1 : \Theta(\tilde{x},1) = 0$$

From Eq. (3.17), it can be seen that the boundary condition $T = T_0$ for $x \rightarrow -\infty$ has been moved to $x = 0$. This means that the process of heating the wall has no influence on the temperature field for $x < 0$. This shows nicely the parabolic nature of the equation.

In Eq. (3.16) the dimensionless eddy diffusivity for heat $\tilde{\varepsilon}_{hn}$ appears as an additional unknown. This quantity is related to the eddy viscosity $\tilde{\varepsilon}_m$ through the similarity between heat and momentum transfer. This is normally indicated by introducing a turbulent Prandtl number defined by

$$\Pr_t = \frac{\varepsilon_m}{\varepsilon_{hn}} = \frac{\tilde{\varepsilon}_m}{\tilde{\varepsilon}_{hn}} \qquad (3.18)$$

The turbulent Prandtl number would be equal to one, if there would be full similarity between momentum and heat transfer. However, this is not the case. Therefore, the turbulent Prandtl number will, in general, be a function of the Reynolds and molecular Prandtl number as well as of the distance from the wall. In literature, there are many different models for the turbulent Prandtl number. Good reviews on this subject can be found in Reynolds (1975), Jischa (1982), Kays and Crawford (1993) and Kays (1994). For the moment, it is sufficient to know that, for the problems under consideration, $\Pr_t$ is only a function of $\tilde{n}$. In the following sections, a few models for the turbulent Prandtl number $\Pr_t$ are provided, which can be used for the calculations of the heat transfer in pipe and channel flows.

Introducing the turbulent Prandtl number into Eq. (3.16) leads to

$$\tilde{u}(\tilde{n})\frac{\partial \Theta}{\partial \tilde{x}} = \frac{1}{\tilde{r}^F}\frac{\partial}{\partial \tilde{n}}\left[\tilde{r}^F a_2(\tilde{n})\frac{\partial \Theta}{\partial \tilde{n}}\right], \quad a_2(\tilde{n}) = 1 + \frac{\Pr}{\Pr_t}\tilde{\varepsilon}_m \qquad (3.19)$$

The energy equation (3.19) has to be solved together with the associated boundary conditions Eq. (3.17). This can be done, by using the method of separation of variables. Therefore, we assume that the temperature can be described by

$$\Theta_j = \Phi_j(\tilde{n})G_j(\tilde{x}) \qquad (3.20)$$

Introducing Eq. (3.20) into Eq. (3.19) results in

$$\frac{\left(\tilde{r}^F a_2(\tilde{n})\Phi'_j(\tilde{n})\right)'}{\tilde{r}^F \tilde{u}(\tilde{n})\Phi_j(\tilde{n})} = \frac{G'_j(\tilde{x})}{G_j(\tilde{x})} = C_j \qquad (3.21)$$

From this equation, we first solve the ordinary differential equation for the function $G_j(\tilde{x})$:

$$\frac{G_j'(\tilde{x})}{G_j(\tilde{x})} = C_j \tag{3.22}$$

This equation has the general solution

$$G_j(\tilde{x}) = D_j \exp\left(C_j \tilde{x}\right) \tag{3.23}$$

From Eq. (3.23) it is obvious that the unknown constant C_j must be smaller than zero, since, for $C_j = 0$, the dependence of the solution on $\tilde{x}$ would be lost and, for $C_j > 0$, the temperature would tend to infinity for $\tilde{x} \to \infty$. If we introduce $C_j = -\lambda_j^2$ into Eq. (3.23), we obtain

$$G_j(\tilde{x}) = D_j \exp\left(-\lambda_j^2 \tilde{x}\right) \tag{3.24}$$

For the function Φ_j the following ordinary differential equation has to be solved

$$\tilde{r}^F \tilde{u}(\tilde{n}) \lambda_j^2 \Phi_j(\tilde{n}) + \left[\tilde{r}^F a_2(\tilde{n}) \Phi_j'(\tilde{n})\right]' = 0 \tag{3.25}$$

Inserting Eq. (3.20) into the boundary conditions Eq. (3.17) results in

$$\tilde{n} = 0 : \frac{\partial \Theta}{\partial \tilde{n}}\bigg|_{\tilde{n}=0} = 0 \Rightarrow G(\tilde{x})\Phi'(0) = 0 \Rightarrow \Phi'(0) = 0 \tag{3.26}$$

$$\tilde{n} = 1 : \Theta(\tilde{x},1) = 0 \Rightarrow G(\tilde{x})\Phi(1) = 0 \Rightarrow \Phi(1) = 0$$

The differential equation (3.25) has to be solved together with the homogeneous boundary conditions (3.26). As we have seen in Chap. 2 for the elementary examples, non-trivial solutions for this problem exist only for selected values of λ_j^2, which are the eigenvalues of Eq. (3.25). Thus, the problem of solving the differential equation (3.25) with boundary conditions (3.26) has been reduced to an eigenvalue problem and is known in literature as a Sturm-Liouville problem (see for example Reid (1980), Kamke (1983), Collatz (1981), Courant and Hilbert (1991) and Sauer and Szabo (1969)). This sort of self-adjoint eigenvalue problem is very common in a lot of physical applications and has been investigated in great detail by several researchers in the past. In the following sections, a short summary of some of the most important properties of this system is given.

Properties of the Sturm-Liouville System

In order to show the basic properties of the Sturm-Liouville system, we consider the problem given by the Eqs. (3.25-3.26) in the following general form:

$$L\left[\Phi_j\right] = \bar{\lambda}_j M\left[\Phi_j\right]$$

(3.27)

with the boundary conditions for the function Φ_j given by Eq. (3.26). For the eigenvalue problem under consideration, the operators L and M take the following form

$$L\left[\Phi_j\right] = -\frac{d}{d\tilde{n}}\left[\tilde{r}^F a_2 \frac{d\Phi_j}{d\tilde{n}}\right]$$

(3.28)

$$M\left[\Phi_j\right] = \tilde{r}^F \tilde{u}(\tilde{n})\Phi_j$$

(3.29)

The most important properties of the Sturm-Liouville system are:

1. The Problem is Self-Adjoint and Positive Definite

In order to show that the eigenvalue problem under consideration (Eqs. (3.25-3.26)). is self-adjoint and positive definite, let us define the following two inner products:

$$<v,w> = \int_0^1 v\,L[w]\,d\tilde{n}$$

(3.30)

and

$$(v,w) = \int_0^1 v\,M[w]\,d\tilde{n}$$

(3.31)

where v and w are two "compare-functions". These functions are assumed to be continuously differentiable functions of $\tilde{n}$ in the interval [0,1]. They satisfy the boundary conditions of the problem given by Eq. (3.26) without vanishing in the whole interval. The eigenvalue problem is now called self-adjoint, if

$$<v,w> = <w,v>, \qquad (v,w) = (w,v)$$

(3.32)

are satisfied. This means that the inner products, defined by Eqs. (3.30-3.31) are symmetric. Additionally, the eigenvalue problem is called positive definite, if

$$<v,v> > 0, \qquad (v,v) > 0$$

(3.33)

Let us first consider the case that the problem is self-adjoint, then

$$<v,w> = -\int_0^1 v\left[\tilde{r}^F a_2(\tilde{n})w'\right]' d\tilde{n} = -\int_0^1 w\left[\tilde{r}^F a_2(\tilde{n})v'\right]' d\tilde{n} = <w,v>$$

(3.34)

must be satisfied. Rearranging Eq. (3.34) gives

$$\int_0^1 \left\{ v\left[\tilde{r}^F a_2(\tilde{n})w'\right]' - w\left[\tilde{r}^F a_2(\tilde{n})v'\right]' \right\} d\tilde{n} = 0 \tag{3.35}$$

The expression in the integral in Eq. (3.35) can be rewritten according to

$$v\left[\tilde{r}^F a_2(\tilde{n})w'\right]' - w\left[\tilde{r}^F a_2(\tilde{n})v'\right]' = \left[v\left(\tilde{r}^F a_2 w'\right) - w\left(\tilde{r}^F a_2 v'\right)\right]' \tag{3.36}$$

Inserting Eq. (3.36) into Eq. (3.35) and evaluating the inegral results in

$$0 = v\left(\tilde{r}^F a_2 w'\right)\Big|_0^1 - w\left(\tilde{r}^F a_2 v'\right)\Big|_0^1 = vw'\Big|^1 - vw'\Big|_0 - wv'\Big|^1 + wv'\Big|_0 \tag{3.37}$$

From the given boundary conditions, Eq. (3.26), we obtain

$$\tilde{n} = 0 : \Phi'_j = 0 \quad \Rightarrow \quad v'(0) = 0, w'(0) = 0 \tag{3.38}$$
$$\tilde{n} = 1 : \Phi_j = 1 \quad \Rightarrow \quad v(1) = 0, \quad w(1) = 0$$

From Eq. (3.38), we see that Eq. (3.37) is identically zero, thus showing that $< v, w > = < w, v >$ is satisfied.

The second requirement is that $(v, w) = (w, v)$. This is also satisfied as

$$(v, w) = \int_0^1 v\left(\tilde{r}^F \tilde{u}w\right) d\tilde{n} = \int_0^1 w\left(\tilde{r}^F \tilde{u}v\right) d\tilde{n} = (w, v) \tag{3.39}$$

From the preceding analysis, it has been shown that the eigenvalue problem under consideration is self-adjoint. We want now to show that the problem is also positive definite. Therefore, we have to prove that

$$(v, v) = \int_0^1 v\left(\tilde{r}^F \tilde{u}v\right) d\tilde{n} = \int_0^1 v^2\left(\tilde{r}^F \tilde{u}\right) d\tilde{n} > 0 \tag{3.40}$$

This is easy to show, since the expression $\left(\tilde{r}^F \tilde{u}\right)$ is a function ≥ 0 in the interval under consideration. For the second inner product, we have to show that

$$< v, v > = -\int_0^1 v\left[\tilde{r}^F a_2 v'\right]' d\tilde{n} > 0 \tag{3.41}$$

This can be shown by partial integration of Eq. (3.41). From

$$-\int_0^1 \underbrace{v}_{a} \underbrace{\left[\tilde{r}^F a_2 v'\right]'}_{b'} d\tilde{n} - \int_0^1 \underbrace{v'}_{a'} \underbrace{\left[\tilde{r}^F a_2 v'\right]}_{b} d\tilde{n} = -v\left[\tilde{r}^F a_2 v'\right]\Big|_0^1 \tag{3.42}$$

one obtains

$$< v, v >= \int_0^1 \tilde{r}^F a_2 (v')^2 \, d\tilde{n} > 0 \tag{3.43}$$

From Eq. (3.43) it is obvious that the expression is larger than zero because $\tilde{r}^F a_2 (\tilde{n})$ is always ≥ 0. Therefore, it has been shown that the eigenvalue problem under consideration is self-adjoint and positive definite. For other wall boundary conditions (constant wall heat flux or boundary conditions of the third kind), the proof that the eigenvalue problem is self-adjoint and positive definite is analogous.

2. Eigenvalues and Eigenfunctions

Having shown that the eigenvalue problem under consideration is self-adjoint, it is relatively simple to prove that the resulting eigenfunctions are orthogonal with respect to a weighting function. In order to show this, the eigenvalue problem (3.27) for two different eigenfunctions Φ_i and Φ_j is considered

$$L\left[\Phi_j\right] = \bar{\lambda}_j M\left[\Phi_j\right] \tag{3.44}$$
$$L\left[\Phi_i\right] = \bar{\lambda}_i M\left[\Phi_i\right]$$

Multiplying the first equation by Φ_i and the second by Φ_j gives, after subtraction

$$\Phi_i L\left[\Phi_j\right] - \Phi_j L\left[\Phi_i\right] = \bar{\lambda}_j \Phi_i M\left[\Phi_j\right] - \bar{\lambda}_i \Phi_j M\left[\Phi_i\right] \tag{3.45}$$

Integrating both sides of Eq. (3.45) between zero and one gives

$$< \Phi_j, \Phi_i > - < \Phi_i, \Phi_j >= \left(\bar{\lambda}_i - \bar{\lambda}_j\right) \int_0^1 \Phi_i M\left[\Phi_j\right] d\tilde{n} \tag{3.46}$$

Because the problem under consideration is self-adjoint, the left hand side of Eq. (3.46) is identical to zero. Thus

$$\left(\bar{\lambda}_i - \bar{\lambda}_j\right) \int_0^1 \Phi_i M\left[\Phi_j\right] d\tilde{n} = 0 \tag{3.47}$$

From this equation, it is obvious that for $i = j$, the left hand side of Eq. (3.47) is identical to zero, because the eigenvalues are the same. For $i \neq j$, the integral $\int_0^1 \Phi_i M\left[\Phi_j\right] d\tilde{n}$ has to be zero in order to fulfill Eq. (3.47). Inserting the definition of the operator M into Eq. (3.47) results in

$$\int_0^1 \tilde{r}^F \tilde{u}(\tilde{n}) \Phi_i \Phi_j \, d\tilde{n} = 0 \quad \text{for} \quad i \neq j \tag{3.48}$$

This means that the eigenfunctions are a set of orthogonal functions in the interval $[0,1]$ with respect to the weighting function $g(\tilde{n})$ defined by

$$g(\tilde{n}) = \tilde{r}^F \tilde{u}(\tilde{n}) \tag{3.49}$$

Eq. (3.48) is an important result for the solution of the partial differential equation, as it can be used later to expand an arbitrary function in terms of eigenfunctions.

3. The Eigenvalues of the Sturm-Liouville System

Normally, the eigenvalues of Eq. (3.27) can only be obtained numerically. Therefore, it is important to find out if we can restrict in advance the search interval for these values. Let us assume that one eigenvalue is complex $\bar{\lambda}_j = \alpha + i\beta$ with the eigenfunction $\Phi_j = a + ib$. Then its complex conjugate $\bar{\lambda}_i = \alpha - i\beta$ is also an eigenvalue with the eigenfunction $\Phi_i = a - ib$. If we introduce these values into Eq. (3.47), one obtains

$$2i\beta \int_0^1 \tilde{r}^F \tilde{u}\left(a^2 + b^2\right) d\tilde{n} = 0 \tag{3.50}$$

Since the weighting function $g(\tilde{n}) = \tilde{r}^F \tilde{u}(\tilde{n})$ is larger or equal to zero in the investigated interval, the integral in Eq. (3.50) must be bigger than zero. Therefore, in order to satisfy relation (3.50), β must be equal to zero. This shows that the eigenvalue problem has only real eigenvalue. Since the eigenvalue problem is also positive definite, it can be shown (see Collatz (1981)) that all eigenvalues are larger or equal to zero. This shows the correctness of our previous assumption $C_j = -\lambda_j^2$, which was based only on physical reasoning.

4. Eigenfunction Expansions for an Arbitrary Function

We have already shown that the eigenfunctions form an orthogonal set of functions. Now we want to prove that an arbitrary "well behaved" function $f(\tilde{n})$ can be represented by an infinite number of eigenfunctions. We assume, therefore, that

$$f(\tilde{n}) = \sum_{j=0}^{\infty} A_j \Phi_j(\tilde{n}) \tag{3.51}$$

is an uniformly convergent series. Multiplying both sides of Eq. (3.51) by $g(\tilde{n})\Phi_i(\tilde{n})$ and integrating over $\tilde{n}$ between zero and one leads to

$$\int_0^1 f(\tilde{n}) g(\tilde{n}) \Phi_i(\tilde{n}) d\tilde{n} = \int_0^1 \sum_{j=0}^{\infty} A_j g(\tilde{n}) \Phi_i \Phi_j d\tilde{n} \tag{3.52}$$

Interchanging on the right hand side of the equation summation and integration results in

$$\int_0^1 f(\tilde{n})g(\tilde{n})\Phi_i(\tilde{n})d\tilde{n} = \sum_{j=0}^{\infty} A_j \int_0^1 g(\tilde{n})\Phi_i\Phi_j d\tilde{n} \tag{3.53}$$

From Eq. (3.53) and Eq. (3.48) it can be seen that the integral on the right hand side is equal to zero if $i \neq j$. Therefore, the constants A_j are given by

$$A_j = \frac{\int_0^1 g(\tilde{n})f(\tilde{n})\Phi_j(\tilde{n})d\tilde{n}}{\int_0^1 g(\tilde{n})\Phi_j^2(\tilde{n})d\tilde{n}} \tag{3.54}$$

The present investigation is intended only to show some of the important properties of the Sturm-Liouville system. The reader is referred to Coddington and Levinson (1955), Sagan (1989), Reid (1980), Kamke (1977) and Myint-U and Debnath (1987) for a more detailed and rigorous mathematical treatment of the Sturm-Liouville eigenvalue problems.

From the previous analysis, it is clear that the solution of the partial differential equation (3.19) can be constructed by linear superposition of solutions given by Eq. (3.20). This results in the following expression for the temperature distribution in the fluid

$$\Theta = \sum_{j=0}^{\infty} A_j \Phi_j(\tilde{n})\exp\left(-\lambda_j^2 \tilde{x}\right) \tag{3.55}$$

The constants A_j in Eq. (3.55) can be obtained by satisfying the boundary condition for $\tilde{x} = 0$. Here we have

$$1 = \sum_{j=0}^{\infty} A_j \Phi_j(\tilde{n}) \tag{3.56}$$

From Eq. (3.54) it is obvious that the constants A_j in Eq. (3.56) are given by

$$A_j = \frac{\int_0^1 \tilde{r}^F \tilde{u}\Phi_j d\tilde{n}}{\int_0^1 \tilde{r}^F \tilde{u}\Phi_j^2 d\tilde{n}} \tag{3.57}$$

The numerator of Eq. (3.57) can be further simplified, by using the differential equation for the eigenvalue system given by Eq. (3.25). Replacing the expression in the integral by Eq. (3.25) leads to

$$\int_0^1 \tilde{r}^F \tilde{u} \Phi_{,j} \, d\tilde{n} = \int_0^1 -\frac{1}{\lambda_j^2} \left[\tilde{r}^F a_2(\tilde{n}) \Phi'_j(\tilde{n}) \right]' d\tilde{n} = \tag{3.58}$$

$$= -\frac{1}{\lambda_j^2} \left[\tilde{r}^F a_2 \Phi'_j \right]\Big|_0^1 = -\frac{1}{\lambda_j^2} \Phi'_j(1)$$

where $a_2(1) = 1$ has been used. Thus, the constants A_j are finally given by

$$A_j = \frac{-\Phi'_j(1)}{\lambda_j^2 \int_0^1 \tilde{r}^F \tilde{u} \Phi_j \, d\tilde{n}} \tag{3.59}$$

Combining Eqs. (3.55) and (3.59), the temperature distribution in the fluid is completely known.

In Chap. 2, we have seen how the eigenvalues and eigenfunctions can be predicted analytically. However, for most technical relevant cases, the eigenvalues and eigenfunctions can only be predicted numerically. In Appendix C a FORTRAN program is described which can be used for this purpose. However, it should be noted, that larger eigenvalues ($j > 10$) might be predictable by asymptotic formulas. This is explained in detail in Chap. 4.

After the temperature distribution in the fluid is known, the Nusselt number can be calculated. The Nusselt number is defined as:

$$\text{Nu}_D = \frac{hD}{k} \tag{3.60}$$

In Eq. (3.60), the Nusselt number depends on the heat transfer coefficient h and the hydraulic diameter D. The hydraulic diameter D is defined as

$$D = \frac{4A}{U} \tag{3.61}$$

where A is the cross sectional flow area of the duct and U is the wetted perimeter. For a circular pipe, one obtains $D = 2R$. For a rectangular channel, the geometrical parameters are shown in Fig. 3.4.

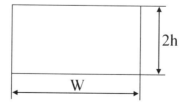

Fig. 3.4: Geometry of a rectangular channel

The channel has width W and height $2h$. Following Eq. (3.61), the hydraulic diameter is equal to

$$D = \frac{4W(2h)}{4h+2W} = \frac{4h}{\frac{2h}{W}+1} \tag{3.62}$$

For the here considered case of a two-dimensional channel ($h/W \to 0$), Eq. (3.62) leads to $D = 4\,h$.

The heat transfer coefficient h, which appears in Eq. (3.60) can be obtained from an energy balance at the surface where the heat transfer process occurs. Here we have

$$h(T_W - T_b) = -k\left.\frac{\partial T}{\partial n}\right|_{n=L} \tag{3.63}$$

Replacing the heat transfer coefficient h in Eq. (3.60) with Eq. (3.63), we obtain the following definition for the Nusselt number

$$\mathrm{Nu}_D = \frac{D\left.\dfrac{\partial T}{\partial n}\right|_{n=L}}{T_W - T_b} \tag{3.64}$$

From this definition, it is obvious that the Nusselt number is a dimensionless temperature gradient at the wall. In Eq. (3.64), T_W denotes the wall temperature, while T_b is the "bulk-temperature" defined by

$$T_b = \frac{\displaystyle\int_0^L uTr^F\,dn}{\displaystyle\int_0^L ur^F\,dn} \tag{3.65}$$

The bulk-temperature is a mass averaged fluid temperature. Introducing the dimensionless quantities defined by Eq. (3.11), we obtain the Nusselt number and the bulk-temperature in dimensionless form

$$\mathrm{Nu}_D = \frac{-4\left.\dfrac{\partial \Theta}{\partial \tilde{n}}\right|_{\tilde{n}=1}}{2^F \Theta_b} \tag{3.66}$$

$$\Theta_b = 2^F \int_0^1 \tilde{u}\tilde{r}^F \Theta\,d\tilde{n} \tag{3.67}$$

Introducing now the known temperature distribution in the fluid (Eq. (3.55)) results in the following expression for the bulk-temperature

$$\Theta_b = -2^F \sum_{j=0}^{\infty} A_j \frac{\Phi'_j(1)}{\lambda_j^2} \exp\left(-\lambda_j^2 \tilde{x}\right) \tag{3.68}$$

and for the Nusselt number

$$\mathrm{Nu}_D = \frac{4\sum_{j=0}^{\infty} A_j \Phi'_j(1) \exp\left(-\lambda_j^2 \tilde{x}\right)}{4^F \sum_{j=0}^{\infty} A_j \frac{\Phi'_j(1)}{\lambda_j^2} \exp\left(-\lambda_j^2 \tilde{x}\right)} \tag{3.69}$$

where $F = 0$ denotes the heat transfer in a planar channel whereas $F = 1$ denotes the heat transfer in a circular pipe. If we are only interested in the Nusselt number for the hydrodynamic and thermal fully developed flow, we can obtain the Nusselt number for the fully developed flow from Eq. (3.69) by setting $\tilde{x} \to \infty$. For this limit only the first term of the sums has to be retained because of the very fast growing eigenvalues with increasing j. One obtains therefore the very simple expression

$$\lim_{\tilde{x} \to \infty} \mathrm{Nu}_D = \mathrm{Nu}_\infty = \frac{4}{4^F} \lambda_0^2 \tag{3.70}$$

We have obtained the result that, in case of a fully developed flow, the Nusselt number (for the constant wall temperature boundary condition) depends solely on the first eigenvalue. This result is important for turbulent flows where the thermal entrance length is normally quite short due to the rapid mixing in the flow.

Laminar Flows

As stated earlier, the heat transfer process, in hydrodynamically fully developed laminar flow in a planar channel and in a pipe, has been extensively studied in literature (see Shah and London (1978)). Therefore, only some results are reported here which help to understand the general behaviour of the heat transfer process in laminar duct flows. One of the most important results for this type of problems is that the temperature distribution as well as the distribution of Nusselt number depend only on the dimensionless axial coordinate

$$\tilde{x} = \frac{x}{L} \frac{1}{\mathrm{Re}_L \mathrm{Pr}} \tag{3.71}$$

and not explicitly on the Reynolds- or Prandtl number (for a pipe flow $L = R$ and for a flow in a planar channel $L = h$). This can easily be understood, if we examine Eq. (3.19). For laminar flows, the velocity distribution is only a function of $\tilde{n}$ and not a function of the Reynolds number (see Eqs. (3.1-3.2)). Additionally, the function $a_2(\tilde{n}) = 1$. Therefore, the problem given by Eq. (3.19) and Eq. (3.17) does not depend on the Reynolds- and the Prandtl number. This fact leads to great simplifications for these type of problems, because the eigenvalues and eigenfunctions have only to be calculated once. The qualitative distribution of the bulk-temperature is displayed in Fig. 3.5, whereas the qualitative distribution of the Nusselt number is given in Fig. 3.6. From Fig. 3.5, it can be seen that the bulk-temperature at the inlet of the duct is equal to the uniform inlet temperature T_0. This is obvious from the definition of T_b, given by Eq. (3.65). With increasing values of $\tilde{x}$, the bulk-temperature attains asymptotically the uniform wall temperature T_W. The distribution of the Nusselt number is shown in Fig. 3.6. Here it can be seen that the value of the Nusselt number is extremely high for very small values of the $\tilde{x}$ coordinate. This is caused by the temperature jump in the wall temperature at the inlet of the duct, which leads to the development of the temperature boundary layer at $\tilde{x} = 0$. With increasing values of the axial coordinate, the Nusselt number decreases and reaches an asymptotic constant value. The value of this constant,

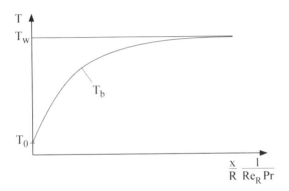

Fig. 3.5: Distribution of the bulk-temperature for a laminar, hydrodynamically fully developed pipe flow with constant wall temperature

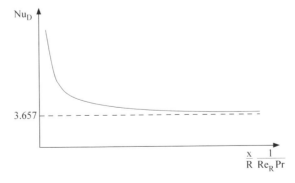

Fig. 3.6: Distribution of the Nusselt number for a laminar, hydrodynamically fully developed pipe flow with constant wall temperature

however, depends on the geometry and applied wall boundary conditions. For the present case of constant wall temperature, the value for the fully developed Nusselt number is

$$Nu_\infty = 3.657 \quad \text{pipe} \tag{3.72}$$

$$Nu_\infty = 7.5407 \quad \text{planar channel}$$

Having shown that, for hydrodynamically fully developed flows, both the temperature field and the distribution of the Nusselt number depend only on the axial coordinate $\tilde{x}$, simple correlations can be obtained quite easily for calculating the Nusselt number and, thus, the heat transfer in ducts for laminar flow. For example, the distribution of the Nusselt number in a pipe for the thermal entrance region can be approximated by (see Shah and London (1978)):

$$Nu_D = \begin{cases} \dfrac{1.077}{(x^*)^{1/3}} - 0.7 & , x^* \le 0.01 \\[4mm] 3.657 + \dfrac{6.874}{\left(10^3 x^*\right)^{0.488}} \exp(-57.2\, x^*) & , x^* > 0.01 \end{cases} \tag{3.73}$$

where a slightly modified axial coordinate has been used:

$$x^* = \frac{x}{D} \frac{1}{Re_D\, Pr} \tag{3.74}$$

D is the hydraulic diameter ($D = 2R$ for the pipe and $D = 4h$ for the planar channel). For the heat transfer in a planar duct similar expressions can be obtained (see Shah and London (1978)):

$$Nu_D = \begin{cases} \dfrac{1.233}{(x^*)^{1/3}} - 0.4 & , x^* \le 0.001 \\[4mm] 7.541 + \dfrac{6.874}{\left(10^3 x^*\right)^{0.488}} \exp(-245x^*) & , x^* > 0.001 \end{cases} \tag{3.75}$$

It is interesting to note that the dependence on the axial coordinate is quite similar for both solutions (Eqs. (3.73) and (3.75)). Only the matching constants differ in order to approximate the heat transfer in a pipe or in a planar channel.

Turbulent Flows

Heat transfer in turbulent duct flows is of greater technical relevance than laminar flow. In turbulent flows, the mixing process is much more effective than in laminar flows. This produces a shorter hydrodynamic entrance length compared to laminar duct flows, while the velocity profiles are much flatter. Thanks to the shorter hydrodynamic entrance length, the flow can be described for a lot of applications either as hydrodynamically fully developed or as hydrodynamically and thermally fully developed. Exceptions are represented by turbulent flows of liquid metals. Due to the very low Prandtl number for these types of applications, the heat transfer behavior of such flows might be more similar to laminar flows.

Before discussing some heat transfer results, the turbulent Prandtl number ($Pr_t = \varepsilon_m / \varepsilon_h$) has to be specified. As it can be seen in Eq. (3.19), the turbulent Prandtl number will significantly influence the term $a_2(\tilde{n})$. If we assume that the transfer of momentum and heat is similar we can set $Pr_t = 1$. Unfortunately, this simple approximation is not realistic for many applications. In literature, a lot of different approaches are described to model the turbulent Prandtl number. The reader is referred to Reynolds (1975), Jischa (1982), Kays and Crawford (1993) and Kays (1994) for good reviews and comparisons between different models. In general it can be noted that

$$Pr_t \rightarrow \infty \qquad \text{for } Re_D \neq 0 \quad \text{and } Pr \rightarrow 0 \tag{3.76}$$
$$Pr_t \rightarrow const. \quad \text{for } Re_D \rightarrow \infty \text{ and } Pr \geq 1$$

This can be explained in the following way: for Pr $\ll 1$ an eddy has a very high molecular heat conduction. Therefore, such an eddy exchanges much more rapidly heat than momentum. This means, that the turbulent Prandtl number increases with decreasing molecular Prandtl number. However, this result is also dependent on the Reynolds number. For increasing Reynolds numbers, the turbulent mixing increases and the influence of molecular heat conduction decreases. This means that, for increasing Reynolds numbers and Prandtl numbers larger than one, the turbulent Prandtl number approaches a constant value. On the other hand, experimental and DNS (direct numerical simulation) results show that the turbulent Prandtl number reaches values larger than one close to the wall. These results show nicely that the heat and momentum transfer differ in proximity of a solid wall.

There are only few reliable measurement data of the turbulent Prandtl number reported in literature (see Jischa (1982), Kays (1994)). The scatter of these measurements is moderate for large molecular Prandtl numbers (Pr > 0.2) and relatively large for lower molecular Prandtl numbers (liquid metal flows). Note that the models presented here for the turbulent Prandtl number can also be applied for liquid metal flows.

Many of the models for the turbulent Prandtl number are based on the Prandtl mixing length concept. Cebeci and Bradshaw (1984), for example, developed a model for the turbulent Prandtl number given by:

$$Pr_t = \frac{\kappa}{\kappa_h} \frac{1 - \exp\left(-y^+ / A^+\right)}{1 - \exp\left(-y^+ / B^+\right)}, \qquad y^+ = \frac{y_W \, u_\tau}{\nu} \tag{3.77}$$

with the quantities

$$\kappa = 0.4, \quad \kappa_h = 0.44, \quad A^+ = 26 \tag{3.78}$$

$$B^+ = \frac{1}{\sqrt{Pr}} \left(\begin{array}{l} 34.96 + 28.79 \log_{10} Pr + 33.95 \left(\log_{10} Pr \right)^2 + \\ +6.33 \left(\log_{10} Pr \right)^3 - 1.186 \left(\log_{10} Pr \right)^4 \end{array} \right)$$

Kays and Crawford (1993) derived a model for the turbulent Prandtl number , given by

$$Pr_t = \left(\frac{1}{2 Pr_{t\infty}} + C Pe_t \sqrt{\frac{1}{Pr_{t\infty}}} - \left(C Pe_t \right)^2 \left[1 - \exp \left(-\frac{1}{C Pe_t \sqrt{Pr_{t\infty}}} \right) \right] \right)^{-1} \tag{3.79}$$

where $Pe_t = \tilde{\varepsilon}_m Pr$, $C = 0.3$ and $Pr_{t\infty} = 0.85$.

Jischa and Rieke (1979) derived a model from modelled transport equations. As a result of their analysis, they obtained the following simple formulation for the turbulent Prandtl number

$$Pr_t = 0.9 + \frac{182.4}{Pr \, Re_D^{0.888}} \tag{3.80}$$

The dependence of the turbulent Prandtl number on the wall coordinate has not been incorporated by Jischa and Rieke (1979).

   One weakness of the model by Kays and Crawford (1993) is that the value for $Pr_{t\infty}$ is a constant for all molecular Prandtl numbers. Using the functional dependence of the turbulent Prandtl number, given in Eq. (3.80), the value of $Pr_{t\infty}$ in Eq. (3.79) can be improved. Following the work of Weigand et al. (1997a), the model given by Eq. (3.79) can be modified by using the following description for $Pr_{t\infty}$

$$Pr_{t\infty} = 0.85 + \frac{100}{Pr \, Re_D^{0.888}} \tag{3.81}$$

By using the above given values for $Pr_{t\infty}$ in Eq. (3.79), the model for the turbulent Prandtl number is able to predict quite reliably the heat transfer for a lot of different applications and a broad range of molecular Prandtl numbers (see Weigand et al. (1997) and Kays and Crawford (1993)).

   Another quite popular model is the one by Azer and Chao (1960). The original expressions for the turbulent Prandtl number given by Azer and Chao (1960) are very complicated. However, the authors reported in their paper the following approximation for $Pr \ll 1$:

$$Pr_t = \frac{1 + 380 / Pe_D^{0.58} \exp \left(-\left(1 - \tilde{n} \right)^{0.25} \right)}{1 + 135 / Re_D^{0.45} \exp \left(-\left(1 - \tilde{n} \right)^{0.25} \right)} \tag{3.82}$$

and for $0.6 \le Pr \le 15$

$$Pr_t = \frac{1+57\,Re_D^{0.46}\,Pr^{0.58}\,\exp\left(-(1-\tilde{n})^{0.25}\right)}{1+135/Re_D^{0.45}\,\exp\left(-(1-\tilde{n})^{0.25}\right)} \qquad (3.83)$$

Additionally, there are a lot of models for the turbulent Prandtl number reported in literature, which are mainly empirical or simply based on measurements. For example, for air flow, the measurements of Ludwieg (1956) can be approximated by (see Reich and Beer (1989))

$$Pr_t = \left(1.53 - 2.82\,\tilde{n}^2 + 3.85\,\tilde{n}^3 - 1.48\,\tilde{n}^4\right)^{-1} \qquad (3.84)$$

After this short review of models for the turbulent Prandtl number we now want to proceed with the discussion of the turbulent heat transfer in a pipe.

Heat Transfer in Turbulent Pipe Flow

If we set the flow index $F = 1$ in the above equations, the equations describing the heat transfer in a hydrodynamically fully developed pipe flow are obtained. This problem has been investigated first by Latzko (1921), who simplified the analysis by putting $Pr_t = 1$ in the energy equation (3.19). After neglecting the molecular heat conduction in Eq. (3.19), Latzko was able to predict the first three eigenvalues for this proplem. Later, Sleicher and Tribus (1957) were concerned with the turbulent Graetz problem. They calculated the first three eigenvalues for various Prandtl and Reynolds numbers. Notter and Sleicher (1972) investigated the problem in great detail. They used an universal turbulent velocity profile for their calculations. For the turbulent Prandtl number, an empirical model was used. For $Pr < 1$, they applied

$$\frac{1}{Pr_t} = \frac{\left(Pr\,\tilde{\varepsilon}_m\right)^{0.39}}{1+\left(Pr\,\tilde{\varepsilon}_m\right)^{0.39}}\left(1+\frac{5.64\sqrt{Pr}}{1.39+\sqrt{\tilde{\varepsilon}_m}}\right) \qquad (3.85)$$

For larger molecular Prandtl numbers, the turbulent Prandtl number was specified by

$$\frac{1}{Pr_t} = \frac{1}{\tilde{\varepsilon}_m}\frac{0.0009\left(y^+\right)^3}{\left(1+\left(0.0818\,y^+\right)^2\right)^{1/2}}, \quad 0 \le y^+ < 45 \qquad (3.86)$$

For $y^+ > 45$, Eq. (3.85) was used again.

As stated before, the Nusselt number for fully developed flow depends only on the first eigenvalue, which can be predicted easily from the previously discussed eigenvalue problem. The Nusselt number for fully developed flow depends on the Reynolds number as well as on the Prandtl number. With increasing Reynolds number, the turbulent mixing increases and therefore the Nusselt number increases for a fixed Prandtl number. For a given value of the Reynolds number, an increasing value of the Prandtl number leads also to an increasing Nusselt number. For the fully developed Nusselt number a lot of correlations are reported in literature. One example is (see Cebeci and Bradshaw (1984))

$$Nu_\infty = \frac{Re_D \, Pr \sqrt{c_f/8}}{0.833 \left[5 Pr + 5 \ln(5 Pr + 1) + 2.5 \ln\left(Re_D \sqrt{c_f/8}\Big/60\right) \right]} \tag{3.87}$$

with the friction factor c_f given by

$$c_f = \frac{0.3164}{Re_D^{0.25}} \tag{3.88}$$

The above equation is valid only for $Pr \geq 0.2$. For smaller molecular Prandtl numbers, the correlation

$$Nu_\infty = 4.8 + 0.0156 \, Re_D^{0.85} \, Pr^{0.93} \tag{3.89}$$

can be used (Sleicher and Rouse (1975)). Fig. 3.7 shows a comparison between predicted and measured Nusselt numbers for fully developed liquid metal flow. The measured data are from Sleicher et al. (1973) and Gilliland et al. (1951). For the calculations, the turbulent Prandtl number model by Azer and Chao (1960) and Kays and Crawford (1993) have been used. It can be seen that both models lead to good agreement between measured and calculated values for a broad range of Peclet numbers. The distribution of the Nusselt number in the thermal entrance region for air flow is shown in Fig. 3.8. Here, experimental data by Abbrecht and Churchill (1960) have been compared with calculations of the Nusselt number, using the velocity distribution given in Appendix A and the turbulent Prandtl number model of Sleicher, Eq. (3.85). Very similar results can be obtained using the turbulent Prandtl number concept of Kays and Crawford (1993).

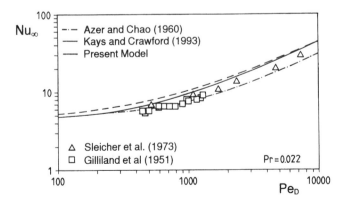

Fig. 3.7: Nusselt number for fully developed pipe flow as a function of the Peclet number (Weigand, Ferguson and Crawford (1997))

As it can be seen, the agreement between calculations and measurements is satisfactory. Fig. 3.9 shows also some results for the thermal entrance region of a liquid metal flow ($Pr = 0.022$) for two different Reynolds numbers. For these calculations, the modified turbulent Prandtl number model of Kays and Crawford (1993) has been used (see Weigand et al. (1997)). As it can be seen, the agreement between calculations and measurement is good. This shows nicely the quality of the used turbulent Prandtl number model, which reproduces the correct behavior in the whole thermal entrance region.

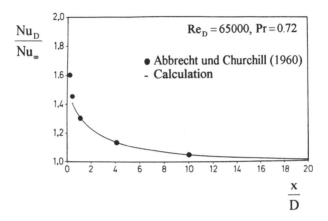

Fig. 3.8: Variation of the Nusselt number in the thermal entrance region of a pipe for $Pr = 0.72$ (Weigand and Beer (1989))

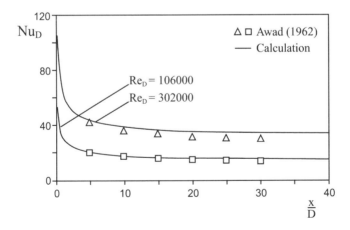

Fig. 3.9: Variation of the Nusselt number in the thermal entrance region of a pipe for $Pr = 0.022$ (Weigand, Ferguson and Crawford (1997))

Turbulent Heat Transfer in a Planar Channel

If we set the flow index $F = 0$ in the above equations, then the equations describing the heat transfer in a hydrodynamically fully developed channel flow are obtained. This problem has been investigated by Sakakibara and Endoh (1977), Shibani and Özisik (1977b) and Özisik et al. (1989). Özisik et al. (1989) pointed out that the eigenvalues in Shibani and Özisik (1977b) have been calculated with a wrong distribution of the wall coordinate y^+ in the turbulence model (a factor of 2 was missing in the definition of y^+). The Nusselt numbers calculated by Özisik et al. (1989) are based on a three layer approximation for the hydrodynamic fully developed velocity distribution (see for details Appendix A, Kays and Crawford (1993), and Reichhardt (1951)). The turbulent eddy viscosity has been calculated according to Spalding (1961) for $y^+ < 40$ and according to Reichhardt for $y^+ \geq 40$. Fig. 3.10 shows the distribution of the Nusselt number in the thermal entrance region of a planar channel for three different Reynolds numbers and for air (Pr = 0.72). The Nusselt number distribution according to Özisik et. al. (1989), which is in good agreement to existing correlations, agrees very well with own predictions using the velocity distribution given in Appendix A .

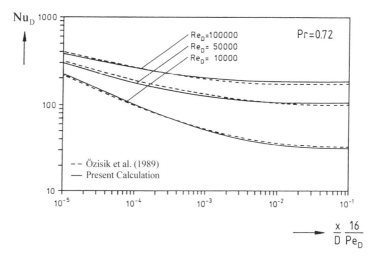

Fig. 3.10: Variation of the Nusselt number in the thermal entrance region of a planar channel for Pr = 0.72

3.2 Thermal Entrance Solutions for an Arbitrary Wall Temperature Distribution

Having obtained the solution for the uniform wall temperature case, the present section deals with the general case of an arbitrary wall temperature distribution.

We assume that the wall temperature changes continuously with $\tilde{x}$ (for $\tilde{x}>0$). Note that if there is a jump in the wall temperature distribution, the solution is then given by Eq. (3.55) and has only to be added to the solution obtained here. The problem under consideration is shown in Fig. 3.11.

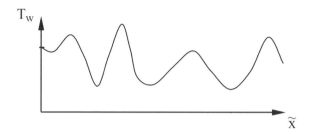

Fig. 3.11: Duct with arbitrary wall temperature

If we introduce a new dimensionless temperature Θ_1, defined by

$$\Theta_1 = \frac{T - T_W(\tilde{x})}{T_0 - T_W(0)} \tag{3.90}$$

the energy equation becomes

$$\tilde{u}(\tilde{n})\frac{\partial \Theta_1}{\partial \tilde{x}} = \frac{1}{\tilde{r}^F}\frac{\partial}{\partial \tilde{n}}\left[\tilde{r}^F\left(1 + Pr\,\tilde{\varepsilon}_{hn}\right)\frac{\partial \Theta_1}{\partial \tilde{n}}\right] + \tilde{u}(\tilde{n})\frac{d\Theta_W}{d\tilde{x}} \tag{3.91}$$

with the boundary conditions

$$\tilde{x} = 0: \Theta_1(0, \tilde{n}) = 1, \text{ temperature jump at } \tilde{x} = 0 \tag{3.92}$$

$$\tilde{n} = 0: \frac{\partial \Theta_1}{\partial \tilde{n}}\bigg|_{\tilde{n}=0} = 0$$

$$\tilde{n} = 1: \Theta_1(\tilde{x}, 1) = 0$$

and the abbreviation $\Theta_W = \left(T_0 - T_W(\tilde{x})\right)/\left(T_0 - T_W(0)\right)$.

As it can be seen from the boundary conditions, Eq. (3.92), and also from Fig. 3.11, the wall temperature distribution has a jump at $\tilde{x} = 0$. Introducing the new defined temperature Θ_1 into the energy equation results in a new term on the right hand side of Eq. (3.91). On the other hand, the boundary conditions for $\tilde{n}=0$ and $\tilde{n}=1$ are homogenous for Θ_1. This means that the solution for the case of an arbitrary wall temperature requires the solution of the non-homogenous equation. In the previous section we have seen that an arbitrary function could be represented by an infinite number of eigenfunctions. Thus, the following expression for the temperature distribution is assumed

$$\Theta_1 = \sum_{m=0}^{\infty} B_m \left(\tilde{x} \right) \Phi_m \left(\tilde{n} \right) \tag{3.93}$$

The eigenfunctions in Eq. (3.93) are already known from the solution of the eigenvalue problem (see Eqs. (3.25-3.26)). In the following analysis, we try to find a
solution of the above non-homogeneous problem by using the orthogonality properties of the eigenfunctions. Multiplying both sides of Eq. (3.93) by
$\tilde{u} \tilde{r}^F \Phi_n \left(\tilde{n} \right)$ and integrating the resulting expressions between 0 and 1 gives the
following equation for the functions B_m

$$B_m = \frac{\int_0^1 \tilde{u} \tilde{r}^F \Phi_m \Theta_1 \, d\tilde{n}}{\int_0^1 \tilde{u} \tilde{r}^F \Phi_m^2 \, d\tilde{n}} = \frac{1}{K_m} \int_0^1 \tilde{u} \tilde{r}^F \Phi_m \Theta_1 \, d\tilde{n} \tag{3.94}$$

Because of the boundary condition for Θ_1 for $\tilde{x} = 0$, the functions B_m have to satisfy the following boundary condition

$$B_m \left(0 \right) = \frac{1}{K_m} \int_0^1 \tilde{u} \tilde{r}^F \Phi_m \Theta_1 \, d\tilde{n} = \frac{-1}{K_m} \int_0^1 1 \cdot \left[\tilde{r}^F a_2 \Phi_m' \right]' d\tilde{n} = \frac{-\Phi_m' \left(1 \right)}{\lambda_m^2 K_m} \tag{3.95}$$

The two other boundary conditions for $\tilde{n} = 0$ and $\tilde{n} = 1$ are automatically satisfied
by Eq. (3.93) because of the employed eigenfunctions.

If we differentiate Eq. (3.94) with respect to $\tilde{x}$, we obtain an ordinary differential equation for the unknown functions B_m

$$\frac{dB_m}{d\tilde{x}} = \frac{1}{K_m} \int_0^1 \tilde{u} \tilde{r}^F \Phi_m \frac{\partial \Theta_1}{\partial \tilde{x}} \, d\tilde{n} \tag{3.96}$$

The expression $\tilde{u} \, \partial \Theta_1 / \partial \tilde{x}$ in the above equation can be replaced by using the energy equation (3.91). This results in

$$\frac{dB_m}{d\tilde{x}} = \frac{1}{K_m} \int_0^1 \frac{\partial}{\partial \tilde{n}} \left[\tilde{r}^F a_2 \left(\tilde{n} \right) \frac{\partial \Theta_1}{\partial \tilde{x}} \right] \Phi_m \, d\tilde{n} + \frac{1}{K_m} \int_0^1 \tilde{u} \tilde{r}^F \Phi_m \frac{d\Theta_W}{d\tilde{x}} \, d\tilde{n} \tag{3.97}$$

The first integral on the right hand side of Eq. (3.97) can be solved by partial integration and taking into account the boundary conditions for Θ_1 and Φ_m. The second integral can be solved directly by using the differential equation for the eigenfunctions, Eq. (3.25). This results in

$$\frac{dB_m}{d\tilde{x}} = -\lambda_m^2 B_m - \frac{\Phi_m' \left(1 \right)}{\lambda_m^2 K_m} \frac{d\Theta_W}{d\tilde{x}} \tag{3.98}$$

The ordinary differential equation (3.98) has the general solution

$$B_m = C_m \exp\left(-\lambda_m^2 \tilde{x}\right) - \frac{\Phi_m'(1)}{\lambda_m^2 K_m} \int_0^{\tilde{x}} \frac{d\Theta_W}{d\tilde{x}} \exp\left(-\lambda_m^2 (\xi - \tilde{x})\right) d\xi \qquad (3.99)$$

The constants C_m can be determined by comparing Eq. (3.99) with Eq. (3.95) for $\tilde{x} = 0$. This results in

$$B_m = -\frac{\Phi_m'(1)}{\lambda_m^2 K_m} \left\{ \exp\left(-\lambda_m^2 \tilde{x}\right) + \int_0^{\tilde{x}} \frac{d\Theta_W}{d\tilde{x}} \exp\left(-\lambda_m^2 (\xi - \tilde{x})\right) d\xi \right\} \qquad (3.100)$$

Introducing Eq. (3.100) into Eq. (3.93) results in the following equation for the temperature distribution

$$\Theta_1 = \Theta(\tilde{x}, \tilde{n}) + \int_0^{\tilde{x}} \Theta\left((\tilde{x} - \xi), \tilde{n}\right) \frac{d\Theta_W}{d\tilde{x}} d\xi \qquad (3.101)$$

In this equation, $\Theta(\tilde{x}, \tilde{n})$ denotes the temperature distribution for the constant wall temperature solution. For $\Theta_W = \mathrm{const.}$ Eq. (3.101) states that $\Theta_1 = \Theta$. From Eq. (3.101) it is obvious, that the temperature distribution for the case of an arbitrary wall temperature can be constructed if the temperature distribution for the uniform wall temperature case is known. This shows very nicely the power of the superposition approach.

It should be noted here, that the solution according to Eq. (3.101) can also be obtained by other methods, for example, by using Duhamel's theorem. The reader is referred to Kays and Crawford (1993) for a different method to derive the above solution for the temperature field for the axial varying wall temperature. The present solution has the advantage that the solution method can also be used in the same way for an arbitrary source term in the energy equation.

3.3 Flow and Heat Transfer in Axially Rotating Pipes with Constant Wall Heat Flux

Fluid flow and heat transfer in rotating systems are not only of considerable theoretical interest, but also of great practical importance. Therefore, transport phenomena in rotating systems have challenged engineers and scientists for a long time. For example, Lord Rayleigh (1917) investigated the dynamics and stability of revolving fluids. Also, some of the classical solutions of the Navier-Stokes equations were obtained for rotating systems (see Schlichting (1982)). Von Karman (1921) investigated the flow induced by a rotating disk and the associated convective heat transfer. The stability of a circular flow in an annulus, formed between two concentric rotating cylinders, was studied by Taylor (1923).

The turbulent flow and heat transfer in an axially rotating pipe is a rather elementary configuration and is therefore suitable as a test case for new turbulence models. Furthermore, many technical applications in rotating machinery are using this simple type of geometry. One of these technical applications is the cooling of gas turbine shafts by means of air flowing through a longitudinal hole in the shaft itself. When a fluid enters a pipe revolving around its axis, tangential forces acting between the rotating pipe and the fluid cause the fluid to rotate with the pipe, resulting in a flow pattern rather different from the one observed in a non-rotating pipe. Rotation was found to have a remarkable influence on the suppression of the turbulent motion because of radially growing centrifugal forces. The effect of pipe rotation on the hydraulic loss has been investigated experimentally by Levy (1929), White (1964) and Shchukin (1967). If the flow is turbulent, the flow is stabilized with increasing rotation rate and the turbulence is suppressed by the centrifugal forces. This can be seen in Fig. 3.12, where flow visualization results from White (1964) are shown. For turbulent flow in a rotating pipe, Borisenko et al. (1973) studied the effect of rotation on the turbulent velocity fluctuations using hot-wire probes and showed that they were suppressed.

$\mathrm{Re}_D = 3520$, $\mathrm{Re}_\varphi = 0$: Turbulent Flow

$\mathrm{Re}_D = 3520$, $\mathrm{Re}_\varphi = 4800$: Laminarized Flow

Fig. 3.12: Flow visualization experiments for the flow in an axially rotating pipe (White (1964))

Murakami and Kikuyama (1980) measured the time-mean velocity components and hydraulic losses in an axially rotating pipe when a fully developed turbulent flow was introduced into the pipe. The results were obtained in dependence of the Reynolds number $\mathrm{Re}_D = \overline{u}\, D / \nu$, the rotation rate $\mathrm{N} = w_W / \overline{u}$ and the length of the rotating pipe. The rotation was found to suppress the turbulence in the flow, and also to reduce the hydraulic loss. With increasing rotational speed, the axial velocity distribution finally approaches the Hagen-Poiseuille flow. Kikuyama et al. (1983) calculated, for the case of a hydrodynamically fully developed flow, the distribution of the axial velocity component by using a simple mixing length approach according to Bradshaw (1969). The distribution of the tangential velocity component was approximated by

$$\frac{w}{w_W} = \left(\frac{r}{R}\right)^2 \tag{3.102}$$

which is in good agreement with experimental data from Murakami and Kikuyama (1980), Kikuyama et al. (1983) and Reich and Beer (1989). Reich and Beer (1989) extended the mixing length approach of Kikuyama et al. (1983) and calculated, for the case of a hydrodynamic and thermal fully developed flow, the axial velocity component and the temperature distribution in the fluid. They showed that in agreement with their measurements, the heat transfer decreased with increasing rotation rate. Hirai and Takagi (1987) and Hirai et al. (1988) calculated the velocity distribution and the heat transfer for the fully developed flow in the axial rotating pipe. In their calculations, it has not been necessary to prescribe the tangential velocity distribution (like Eq. (3.102)) at the beginning of the calculation procedure as it has been done by Kikuyama et al. (1983) and Reich and Beer (1989). However, they obtained the universal distribution of the tangential velocity as a result of their calculations. For the hydrodynamically fully developed flow Weigand and Beer (1989a) solved analytically the energy equation and calculated the local distribution of the Nusselt number for the thermal entrance region. The calculated Nusselt numbers compared very well with the experimentally obtained results of Reich (1988). For the case of a convectively cooled pipe, Weigand and Beer (1992b) obtained an analytical solution for the Nusselt number distribution in the thermal entrance region. It could be shown, that the thermal wall boundary condition influenced the values of the Nusselt number. This is because of the laminarization effect within the pipe. For the case of a hydrodynamically and thermally developed flow, Weigand and Beer (1992a) calculated numerically the distribution of the axial velocity distribution and the local Nusselt number as a function of the axial coordinate. Because of the simple geometry involved, the fully developed flow in an axial rotating pipe has been investigated both by using LES (large eddy simulation) and DNS (direct numerical simulation). Eggels and Nieuwstadt (1993) calculated the axial and tangential velocity distribution in the axial rotating pipe by using LES. Their calculations showed the parabolic distribution of the tangential velocity component and agreed well with measurements of Reich (1988) and Nishibori et al. (1987). Orlandi and Fatica (1997) investigated the flow by DNS for low Reynolds numbers. Malin and Younis (1997) employed the closures for the transport equations of Gibson and Younis (1986) and Speziale et al. (1991) to predict the flow and heat transfer in an axially rotating pipe. The results were compared to experimental data and LES results and good agreement was found for both models. Rinck and Beer (1999) used a modified Reynolds stress model to predict the fully developed flow in an axially rotating pipe. They found good agreement between their predictions and experimental data of Reich (1988). Recently, Speziale et al. (2000) presented a comprehensive analysis of the modeling of turbulent flows in rotating pipes. They discussed also the main physical characteristics of the flow and surveyed the capabilities of different classes of closure models to reproduce them.

3.3.1 Velocity Distribution for the Hydrodynamically Fully Developed Flow in an Axial Rotating Pipe

The hydrodynamically fully developed velocity distribution in an axially rotating pipe can be calculated by using a simple mixing length approach. This procedure is only briefly outlined here, since our main focus is on analytical methods relevant for solving heat transfer problem. A more detailed explanation of the method as well as a review on other methods for calculating the velocity distribution in an axial rotating pipe is given in Appendix B.

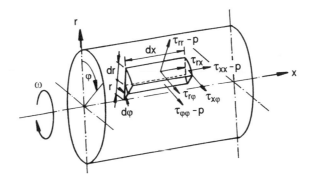

Fig. 3.13: Physical model and coordinate system (Reich and Beer (1989))

Figure 3.13 shows the geometry and the cylindrical coordinate system with the x, r, φ axis and the related velocity components u, v, w.

If we assume a hydrodynamically fully developed turbulent flow with constant fluid properties, the Navier-Stokes equation for the rotational symmetric case are given by (see for example Reich (1988)):

$$-\rho \frac{w^2}{r} = -\frac{\partial p}{\partial r} - \frac{\rho}{r} \frac{\partial}{\partial r}\left(r \overline{v'v'}\right) + \rho \frac{\overline{w'w'}}{r} \tag{3.103}$$

$$0 = \frac{1}{r^2} \frac{\partial}{\partial r}\left(\mu r^3 \frac{\partial}{\partial r}\left(\frac{w}{r}\right) - \rho r^2 \overline{v'w'}\right) \tag{3.104}$$

$$0 = \frac{\partial p}{\partial x} + \frac{1}{r} \frac{\partial}{\partial r}\left(\mu r \frac{\partial u}{\partial r} - \rho r \overline{u'v'}\right) \tag{3.105}$$

with the boundary conditions

$$r = 0: \quad w = 0, \quad v = 0, \quad \frac{\partial u}{\partial r} = 0 \tag{3.106}$$

$$r = R: \quad w = w_W, \quad u = 0, \quad v = 0$$

Since we assume a hydrodynamically fully developed flow, the radial velocity component v is zero everywhere and the continuity equation results in the fact that the axial velocity component u is only a function of r. As mentioned before, experimental data indicate that the tangential velocity distribution is universal and can be approximated by Eq. (3.102). Fig. 3.14 shows the tangential velocity distribution for different rotation rates N as well as for different Reynolds numbers. From Fig. 3.14 it is obvious that Eq. (3.102) is a very good approximation for the tangential velocity distribution. Introducing Eq. (3.102) into Eq. (3.104) shows that $\overline{v'w'} \sim C r$. This relation is of course only correct far away from the wall. However, it is interesting to note that the above given linear relation has been confirmed by DNS calculations.

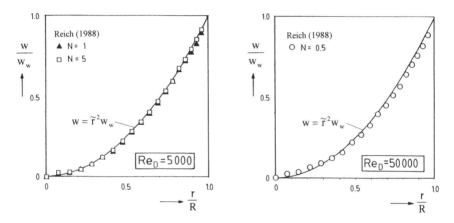

Fig. 3.14: Tangential velocity distribution (Reich and Beer (1989))

The axial velocity distribution $u(r)$ can be calculated from Eq. (3.105). In order to do this, the turbulent shear stress in this equation has to be related to the mean velocity gradients. This can be done by using a mixing length model according to Koosinlin et al. (1975). This results in the following expression for the turbulent shear stress

$$\overline{\rho u' v'} = \rho l^2 \left[\left(\frac{\partial u}{\partial r} \right)^2 + \left(r \frac{\partial}{\partial r} \left(\frac{w}{r} \right) \right)^2 \right]^{1/2} \frac{\partial u}{\partial r} = \varepsilon_m \rho \frac{\partial u}{\partial r} \tag{3.107}$$

where the mixing length distribution l is given by (Reich and Beer (1989))

$$\frac{l}{l_0} = \left(1 - \frac{1}{6}\text{Ri}\right)^2 \tag{3.108}$$

The mixing length distribution l_0 is the one for a non-rotating pipe (see Appendix A, Eq. (A.20)). The Richardson number in Eq. (3.108) describes the effect of pipe rotation on the turbulent motion and is defined by

$$\text{Ri} = \frac{2\dfrac{w}{r}\dfrac{\partial}{\partial r}(wr)}{\left(\dfrac{\partial u}{\partial r}\right)^2 + \left(r\dfrac{\partial}{\partial r}\left(\dfrac{w}{r}\right)\right)^2} \tag{3.109}$$

Without rotation, $\text{Ri} = 0$ and there exists a fully developed turbulent pipe flow. If $\text{Ri} > 0$, i.e. for an axially rotating pipe with a radially growing tangential velocity, the centrifugal forces suppress the turbulent fluctuations and the mixing length decreases. Inserting Eqs. (3.107-3.109) into Eq. (3.105) results in a strongly non-linear ordinary differential equation for the axial velocity component u. This is shown in detail in Appendix B. Fig. 3.15 shows a comparison between calculated and measured axial velocity distributions according to Reich and Beer (1989). It can be seen that the numerical calculations agree well with the measurements.

In Appendix B, an analytical approximation for the velocity distribution in the axial rotating pipe is given (Weigand and Beer (1994)).

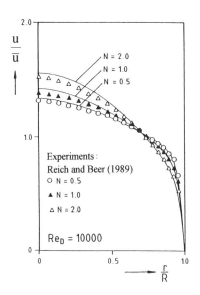
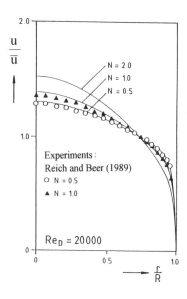

Fig. 3.15: Axial velocity distribution as a function of the rotation rate N (Reich and Beer (1989))

3.3.2 Thermal Entrance Solution for Constant Wall Heat Flux

The thermal entrance solution for the case of a constant heat flux at the wall can also be obtained by the method of separation of variables. However, because of the different wall boundary condition, the approach for solving this problem is different to the case of a constant wall temperature.

If we assume that the flow is incompressible and that the fluid properties are constant, the energy equation for the hydrodynamically fully developed flow, with negligible axial heat conduction effects, takes the form

$$\rho c_p\, u(r)\frac{\partial T}{\partial x} = \frac{1}{r}\frac{\partial}{\partial r}\left[r\left(k\frac{\partial T}{\partial r} - \rho\,\overline{v'T'} \right)\right] \tag{3.110}$$

If we replace the turbulent heat flux in Eq. (3.110) by the eddy viscosity and the turbulent Prandtl number, we obtain

$$\rho c_p u(r)\frac{\partial T}{\partial x} = \frac{1}{r}\frac{\partial}{\partial r}\left[\left\{ k + \rho\, c_p \frac{l^2}{\mathrm{Pr}_t}\left[\left(\frac{\partial u}{\partial r}\right)^2 + \left(r\frac{\partial}{\partial r}\left(\frac{w}{r}\right)\right)^2\right]^{1/2}\right\} r\frac{\partial T}{\partial r}\right] \tag{3.111}$$

with the turbulent Prandtl number for air flow given by

$$\frac{1}{\mathrm{Pr}_t} = 1.53 - 2.82\left(\frac{r}{R}\right)^2 + 3.85\left(\frac{r}{R}\right)^3 - 1.48\left(\frac{r}{R}\right)^4 \tag{3.84}$$

Eq. (3.111) has to be solved with the following boundary conditions

$$x=0:\ T=T_0 \tag{3.112}$$

$$r=0:\frac{\partial T}{\partial r}=0,\qquad r=R:-k\frac{\partial T}{\partial r}=\dot{q}_W=\text{const.}$$

For the constant wall temperature case, the driving temperature potential $T_W - T_0$ can be used to scale the temperature. This is not the case for a constant wall heat flux. Therefore, the following quantities are used to make the equations dimensionless:

$$\Theta = \frac{T-T_0}{(\dot{q}_W R/k)},\ \ \tilde{r}=\frac{r}{R},\ \ \tilde{x}=\frac{x}{R}\frac{2}{\mathrm{Re_D}\,\mathrm{Pr}},\ \ \tilde{u}=\frac{u}{\bar{u}},\ \ \mathrm{Re}_D=\frac{\bar{u}\,D}{v} \tag{3.113}$$

As it can be seen from Eq. (3.113), the temperature is made dimensionless with the heat flux density at the wall, while the axial and radial coordinates are made dimensionless in the usual way. Introducing the dimensionless quantities into the Eqs. (3.111-3.112) results in:

$$\tilde{u}\tilde{r}\frac{\partial\Theta}{\partial\tilde{x}} = \frac{\partial}{\partial\tilde{r}}\left[\tilde{r}\,a_2(\tilde{r})\frac{\partial\Theta}{\partial\tilde{r}}\right] \tag{3.114}$$

with the boundary conditions

$$\tilde{x}=0: \; \Theta=0 \tag{3.115}$$

$$\tilde{r}=0: \frac{\partial \Theta}{\partial \tilde{r}}=0, \qquad \tilde{r}=1: \frac{\partial \Theta}{\partial \tilde{r}}=-1$$

and the function $a_2\left(\tilde{r}\right)$

$$a_2\left(\tilde{r}\right)=1+\frac{\mathrm{Pr}}{\mathrm{Pr}_t}\mathrm{Re}_D\, l^2\left[\left(\frac{\partial \tilde{u}}{\partial \tilde{r}}\right)^2+\left(\frac{1}{2}\tilde{r}\,N\right)^2\right]^{1/2} \tag{3.116}$$

Because the wall boundary condition for $\tilde{r}=1$ is non-homogeneous, the solution for the problem given by Eqs. (3.114-3.115) has to be split into two problems. In the first problem only a particular solution for the energy equation is searched. The second problem can then be solved afterwards by using the method of separation of variables. This can be done, because the energy equation, as well as the wall boundary conditions, are linear.

Introducing the function

$$\Theta=\Theta_1+\Theta_2 \tag{3.117}$$

into the above equations results in the two problems

$$\tilde{u}\tilde{r}\frac{\partial \Theta_1}{\partial \tilde{x}}=\frac{\partial}{\partial \tilde{r}}\left[\tilde{r}\,a_2(\tilde{r})\frac{\partial \Theta_1}{\partial \tilde{r}}\right] \tag{3.118}$$

with the boundary conditions

$$\tilde{r}=0: \frac{\partial \Theta_1}{\partial \tilde{r}}=0, \qquad \tilde{r}=1: \frac{\partial \Theta_1}{\partial \tilde{r}}=-1 \tag{3.119}$$

and

$$\tilde{u}\tilde{r}\frac{\partial \Theta_2}{\partial \tilde{x}}=\frac{\partial}{\partial \tilde{r}}\left[\tilde{r}\,a_2(\tilde{r})\frac{\partial \Theta_2}{\partial \tilde{r}}\right] \tag{3.120}$$

with the boundary conditions

$$\tilde{x}=0: \; \Theta_2=-\Theta_1 \tag{3.121}$$

$$\tilde{r}=0: \frac{\partial \Theta_2}{\partial \tilde{r}}=0, \qquad \tilde{r}=1: \frac{\partial \Theta_2}{\partial \tilde{r}}=0$$

Because only one particular solution of the energy equation has to be found by solving the problem for Θ_1, the boundary condition for $\tilde{x}=0$ can be ignored. This implies that, for the second problem, the boundary condition for $\tilde{x}=0$ has to be modified. Physically, the solution Θ_1 represents the temperature distribution for the fully developed flow, i.e. far away from the pipe entrance, where the fluid temperature monotonously increases due to the heat addition through the wall.

Temperature Distribution for the Fully Developed Flow (Θ_1)

For the fully developed flow, the temperature distribution in the pipe increases linearly with growing values of the axial coordinate. This can be shown by looking at an energy balance for the pipe section shown in Fig. 3.16.
The energy balance leads to

$$2\pi R\, x\, \dot{q}_W = \int_0^{2\pi} \int_0^R \rho c_p (T - T_0) u r\, dr\, d\varphi \tag{3.122}$$

Introducing the dimensionless quantities from Eq. (3.113) into Eq. (3.122) gives

$$\tilde{x} = \int_0^1 \Theta_1 \tilde{u} \tilde{r}\, d\tilde{r} \tag{3.123}$$

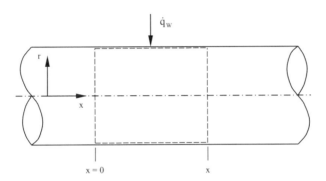

Fig. 3.16: Energy balance for the pipe with an uniform heat flux at the wall

From Eq. (3.123), it is obvious that the temperature in the fluid increases linearly with growing axial distances for the fully developed flow. This motivates the introduction of the following expression for Θ_1

$$\Theta_1 = C_1 \tilde{x} + \psi(\tilde{r}) \tag{3.124}$$

Introducing Eq. (3.124) into Eqs. (3.118-3.119) and into Eq. (3.123) results in

$$\tilde{u}\tilde{r} C_1 = \frac{d}{d\tilde{r}}\left(a_2(\tilde{r})\tilde{r}\frac{d\psi}{d\tilde{r}} \right) \tag{3.125}$$

$$\tilde{r} = 0 : \frac{d\psi}{d\tilde{r}} = 0, \quad \tilde{r} = 1 : \frac{d\psi}{d\tilde{r}} = -1 \tag{3.126}$$

$$\tilde{x} = \int_0^1 \tilde{x} \, C_1 \tilde{u} \tilde{r} d\tilde{r} + \int_0^1 \psi(\tilde{r}) \tilde{u} \tilde{r} d\tilde{r} \tag{3.127}$$

From Eq. (3.127), it can be seen that the unknown constant C_1 is determined by the integral energy balance. Solving the above set of ordinary differential equations results in the following temperature distribution for the fully developed flow

$$\Theta_1(\tilde{r}, \tilde{x}) = -2\tilde{x} + \bar{\psi}(\tilde{r}) + C_2 \tag{3.128}$$

with the abbreviations

$$\bar{\psi}(\tilde{r}) = \int_0^{\tilde{r}} \left(\frac{-2}{\eta \, a_2(\eta)} \int_0^\eta \tilde{u}(\xi) \xi d\xi \right) d\eta, \quad C_2 = -\int_0^1 2\bar{\psi}(\tilde{r}) \tilde{r} \tilde{u} \, d\tilde{r} \tag{3.129}$$

The Temperature Distribution Θ_2

Introducing the obtained temperature distribution for the fully developed flow into Eq. (3.121) results in the following problem for Θ_2

$$\tilde{u}\tilde{r} \frac{\partial \Theta_2}{\partial \tilde{x}} = \frac{\partial}{\partial \tilde{r}} \left[\tilde{r} \, a_2(\tilde{r}) \frac{\partial \Theta_2}{\partial \tilde{r}} \right] \tag{3.120}$$

$$\tilde{x} = 0 : \ \Theta_2 = \bar{\psi}(\tilde{r}) + C_2 \tag{3.130}$$

$$\tilde{r} = 0 : \frac{\partial \Theta_2}{\partial \tilde{r}} = 0, \qquad \tilde{r} = 1 : \frac{\partial \Theta_2}{\partial \tilde{r}} = 0$$

The latter can be solved using the method of separation of variables. Introducing

$$\Theta_{2j} = F_j(\tilde{x}) \Phi_j(\tilde{r}) \tag{3.131}$$

into Eqs. (3.120, 3.130) results in the following eigenvalue problem

$$\left(a_2(\tilde{r}) \tilde{r} \, \Phi_j' \right)' + \lambda_j^2 \, \tilde{r} \tilde{u} \, \Phi_j' = 0 \tag{3.132}$$

with the boundary conditions

$$\tilde{r} = 0 : \Phi_j' = 0, \ \tilde{r} = 1 : \Phi_j' = 0 \tag{3.133}$$

and an arbitrary normalization condition

$$\tilde{r} = 0 : \Phi_j = 1 \tag{3.134}$$

The functions $F_j(\tilde{x})$ are given by

$$F_j(\tilde{x}) = D_j \exp(-\lambda_j^2 \tilde{x}) \tag{3.135}$$

The eigenfunctions can be predicted with a program similar to the one given in Appendix C. The temperature distribution Θ_2 is then given by

$$\Theta_2 = \sum_{j=0}^{\infty} A_j \Phi_j(\tilde{r}) \exp\left(-\lambda_j^2 \tilde{x}\right) \tag{3.136}$$

The constants A_j can be calculated with a procedure similar to the one employed in section 3.1 for the constant wall temperature case. Using the orthogonality of the eigenfunctions leads for the present case to

$$A_j = \int_0^1 \psi(\tilde{r}) \tilde{u}(\tilde{r}) \tilde{r} \, \Phi_j(\tilde{r}) \, d\tilde{r} \Big/ \int_0^1 \tilde{u}\tilde{r} \, \Phi_j^2(\tilde{r}) d\tilde{r} \tag{3.137}$$

Both the eigenvalues and eigenfunctions can be calculated numerically. Table 2.1 reports some eigenvalues and constants $A_j \Phi_j(1)$ calculated for different Reynolds numbers and rotation rates N. Note that the eigenvalues increase with increasing Reynolds number and decrease with increasing rotation rate. This means that increasing Reynolds numbers result in a better mixing in the flow and lead to shorter thermal entrance lengths, whereas increasing rotation rates lead to laminarization and therefore to a longer thermal entrance length.

Table 2.1. Eigenvalues and constants for different rotation rates and different Reynolds numbers (Pr = 0.71).

Re$_D$ = 5000

N	λ_0^2	λ_1^2	λ_2^2	$A_0\Phi_0(1)$	$A_1\Phi_1(1)$	$A_2\Phi_2(1)$
0	0	152.95	436.18	0	0.0287	0.0156
1	0	80.76	250.74	0	0.0409	0.0183
2	0	46.13	146.25	0	0.0654	0.0259
3	0	30.56	97.52	0	0.0946	0.0358
5	0	18.62	59.73	0	0.1467	0.0537

Re$_D$ = 20000

N	λ_0^2	λ_1^2	λ_2^2	$A_0\Phi_0(1)$	$A_1\Phi_1(1)$	$A_2\Phi_2(1)$
0	0	505.25	1476.01	0	0.0077	0.0044
1	0	216.04	687.15	0	0.0140	0.0058
2	0	108.93	350.64	0	0.0266	0.0010
3	0	63.24	203.82	0	0.0450	0.0165
5	0	28.57	92.04	0	0.0960	0.0349

Re$_D$ = 50000

N	λ_0^2	λ_1^2	λ_2^2	$A_0\Phi_0(1)$	$A_1\Phi_1(1)$	$A_2\Phi_2(1)$
0	0	1115.14	3281.90	0	0.0034	0.0018
1	0	436.34	1396.90	0	0.0068	0.0027
2	0	210.11	676.36	0	0.0136	0.0048
3	0	115.92	375.17	0	0.0246	0.0088
5	0	44.17	142.76	0	0.0630	0.0225

From Table 2.1, an interesting fact can be noticed. For the considered wall boundary conditions, $\lambda_0 = 0$ is a possible eigenvalue. This eigenvalue results in the possible eigenfunction $\Phi_0 = 1$. If we introduce this eigenvalue and the eigenfunction into Eq. (3.137) we obtain $A_0 = 0$ (because the numerator of Eq. (3.137) is zero according to Eq. (3.127)). Therefore, $\lambda_0 = 0$ can be excluded from the considerations and Eq. (3.137) can further be simplified. One finally obtains

$$A_j = \frac{-\Phi_j'(1)/\lambda_j^2}{\int_0^1 \tilde{u}\tilde{r}\,\Phi_j^2(\tilde{r})\,d\tilde{r}}, \quad j=1,2,3\ldots \tag{3.138}$$

After having predicted Θ_1 and Θ_2, the temperature distribution in the fluid is completely known

$$\Theta(\tilde{r},\tilde{x}) = -2\,\tilde{x} + \psi(\tilde{r}) + \sum_{j=1}^{\infty} A_j \Phi_j(\tilde{r})\exp\left(-\lambda_j^2\,\tilde{x}\right) \tag{3.139}$$

With the help of Eq. (3.139), the distribution of the Nusselt number can be obtained

$$\mathrm{Nu}_D = \frac{D\left.\dfrac{\partial T}{\partial \tilde{r}}\right|_{r=R}}{T_W - T_b} = \frac{2}{\overline{\psi}(1) + C_2 + \sum_{j=1}^{\infty} A_j \Phi_j(1)\exp\left(-\lambda_j^2\,\tilde{x}\right)} \tag{3.140}$$

For the hydrodynamically and thermally fully developed flow, the equation for the Nusselt number reduces to

$$\mathrm{Nu}_\infty = \frac{2}{\overline{\psi}(1) + C_2} \tag{3.141}$$

Figure 3.17 shows the Nusselt number for the fully developed flow, for Pr = 0.71, as a function of the rotation rate N and the Reynolds number (Reich (1988), Reich and Beer (1989)). It can be seen that the Nusselt number strongly decreases with increasing rotation rate N, which is caused by flow laminarization. For $N \rightarrow \infty$, the Nusselt number approaches the limit $\mathrm{Nu}_\infty = 4.36$ for laminar pipe flow.

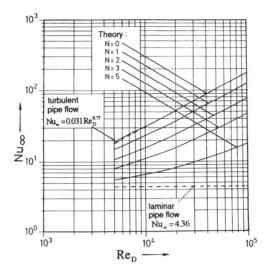

Fig. 3.17: Nusselt number for fully developed flow in an axially rotating pipe (adapted from Reich and Beer (1989), Weigand and Beer (1989a))

Fig. 3.18 shows the distribution of Nu / Nu_∞ in the thermal entrance region for air flow. It can be seen that the length of the thermal entrance region is increased with increasing rotation rates. This is caused by the laminarization of the flow. For N $\rightarrow \infty$, the thermal entrance length approaches the one for laminar pipe flow which is given by

$$\frac{L_{th}}{D} = 0.05 \ Re_D \ Pr \tag{3.142}$$

This effect has interesting implications, specially for larger Reynolds numbers. For example for a Reynolds number of about 50 000, the flow is not even thermally fully developed after 80 pipe diameters for N = 1. This shows very clearly that care has to be taken in predicting the heat transfer for such types of flows.

Reich (1988) obtained experimentally the Nusselt numbers in an axially rotating pipe with constant heat flux at the wall. The rotating test section was 40 diameter in length. He compared his experimental results with calculations for a fully developed flow and found that the difference between experimental data and calculated values increased with growing values of the rotation rate.

However, if one calculates the Nusselt numbers from the above solution for the thermal entrance region and compares them to the measured values, very good agreement is obtained. This is shown in Fig. 3.19 (Weigand and Beer (1989a)).

The results presented here can easily be extended to other wall boundary conditions like constant wall temperature or uniform external heating or cooling. The reader is referred to Weigand and Beer (1992b) for such cases.

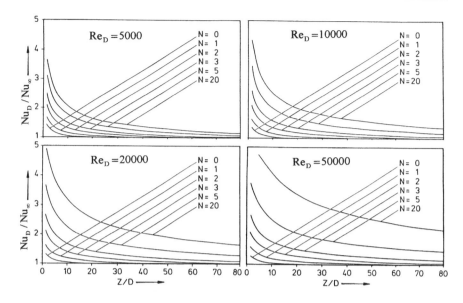

Fig. 3.18: Nusselt number distribution in the thermal entry length for different rotation rates N and Pr = 0.71 (Weigand and Beer (1989a))

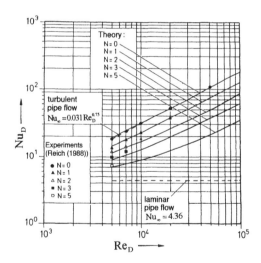

Fig. 3.19: Nusselt numbers at $x/D = 40$ for various values of the rotation rate and Pr = 0.71 (Weigand and Beer (1989a))

The examples of the present chapter have shown that even complicated technical problems can be solved using analytical methods. It is important to note that in the analytical solution the physical significance of individual effects (wall boundary conditions, type of fluid, rotation,…) can be more easily identified than in the numerical solution of the problem.

Problems

3-1. Consider the following eigenvalue problem

$$Y_n'' + \lambda_n^2 Y_n = 0$$

with the boundary conditions

$$Y_n'(0) = Y_n(1) = 0$$

a.) Prove that the eigenvalue problem is self-adjoint and positive definite.
b.) Show that the eigenfunctions form an orthogonal set of functions in the interval [0,1].
c.) Calculate the eigenfunctions and the eigenvalues. Use a normalizing condition that $Y(0) = 1$.

3-2. Show that the following three eigenvalue problems are self-adjoint and positive definite:

$$\left(1 + \tilde{y}^2\right)\lambda_n \Phi_n + \Phi_n'' = 0$$
$$\Phi_n(0) = \Phi_n(1) = 0$$

$$\tilde{r}R_n'' + R_n' + \lambda_n^2 \tilde{r}R_n = 0$$
$$R'(0) = R(1) = 0$$

$$\left(1 - \tilde{x}^2\right)Y_n'' - 2\tilde{x}Y_n' + \lambda_n\left(\lambda_n + 1\right)Y_n = 0$$
$$Y_n(0) = Y_n(1) = 0$$

3-3. Consider the flow and heat transfer is a planar channel with height $2h$. The plates have a constant wall temperature T_0 for $x < 0$ and T_W for $x \geq 0$. The velocity profile of the laminar flow in the planar duct can be simplified to be a slug flow profile with $u = \bar{u} = $ const.. The velocity component in the y – direction $v = 0$. All fluid properties can be considered to be constant. The temperature distribution in the fluid can be calculated from the energy equation

$$u\frac{\partial T}{\partial x} = a\frac{\partial^2 T}{\partial y^2}$$

with the boundary conditions

$$x = 0 : T = T_0$$

$$y = 0 : \frac{\partial T}{\partial y} = 0, \qquad y = h : T = T_w$$

a.) Introduce the following dimensionless quantities

$$\tilde{x} = \frac{x}{h} \frac{a}{\bar{u}h}, \quad \tilde{y} = \frac{y}{h}, \quad \Theta = \frac{T - T_w}{T_0 - T_w}$$

What is the resulting equation and the boundary condition in dimensionless form?

b.) Use the method of separation of variables to solve the given problem ($\Theta = f(\tilde{x})g(\tilde{y})$). What equations are obtained for the functions f and g?

c.) Calculate the eigenfunctions and display the first and the third eigenfunction as a function of $\tilde{y}$ in the interval $[0,1]$.

d.) Predict the complete solution for the temperature field.

3-4. A planar channel with the height $2h$ is subjected to a hydrodynamically fully developed laminar flow. The fluid enters the channel with the constant temperature T_0. The walls of the channel are adiabatic for $x < 0$, whereas for $x \geq 0$ the walls are heated by a constant heat flux $\dot{q}_w$. Assume that all fluid properties are constant. The velocity distribution is given by

$$\frac{u}{\bar{u}} = \frac{3}{2}\left(1 - \left(\frac{y}{h}\right)^2\right)$$

and the Peclet number is large enough, so that axial heat conduction can be ignored in the fluid.

a.) Calculate the temperature distribution in the fluid for large axial values. Display the form of the radial temperature distribution for a given value of the axial coordinate.

b.) Predict the value of the Nusselt number for the fully developed flow.

c.) Predict the complete solution of the problem (**hint:** Split the problem in two separate problems).

3-5. Consider the production process of a thin foil. The foil is moved with constant velocity u_∞ through a thin slit and cools down in the surrounding air. The process is schematically shown in the following figure

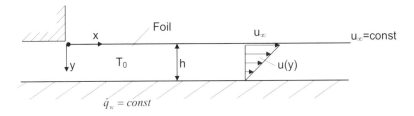

The distance between the foil and the wall is h. The wall is cooled for $x > 0$ and takes out heat from the surrounding air with the constant heat flux $\dot{q}_W$. For large axial values, a hydrodynamically fully developed velocity profile according to

$$\frac{u}{\bar{u}} = 1 - \frac{y}{h}$$

is established. The temperature distribution in the air gap should be predicted for large axial values. The upper wall (foil) can be assumed to be adiabatic and all fluid properties are constant.

a.) Simplify the energy equation for the given problem. Derive the boundary conditions for $y = 0$ and $y = h$ and make the energy equation dimensionless.

b.) Derive an integral energy balance for the thermal fully developed flow. What functional form the temperature distribution must have for the fully developed flow?

c.) Introduce the above derived expression for the temperature distribution into the energy equation and solve the equation with the given boundary conditions.

d.) Calculate the Nusselt number at the lower wall. Use $T(x, y = 0) - T(x, y = h)$ as the driving temperature difference for the Nusselt number.

4 Analytical Solutions for Sturm - Liouville Systems with Large Eigenvalues

The numerical method for obtaining the eigenvalues and eigenfunctions of Eq. (3.25) is explained in Chap. 3 and Appendix C. This numerical procedure gets cumbersome and difficult if the eigenvalues become large, because of the increasing number of zero points of the eigenfunction. This is elucidated in Fig. 4.1, which shows several eigenfunctions for a turbulent pipe flow with constant wall temperature.

As it can be seen from Fig. 4.1, the higher eigenfunctions strongly oscillate. It can be shown, that the eigenfunction of number j has also j zero points. This means that, for larger j, it is difficult to capture exactly the shape of the eigenfunctions by means of a numerical method. An increasingly larger number of grid points is needed to resolve the higher eigenfunctions. Because of this reason, there has been a lot of interest in obtaining analytical approximations of eigenfunctions for larger eigenvalues. Good reviews about this subject can be found in Erdelyi (1956) and in Kamke (1983). For laminar flows, the hydrodynamically fully developed velocity profile is a simple quadratic function of the spatial coordinate. In addition, the function $a_2(\tilde{n})$ in Eq. (3.25) is equal to one. These two simplifications allow that asymptotic approximations can be obtained relatively easily for laminar flow problems. A good review on this topic is given in Shah and London (1978). In the present section, only some of this work will be presented. Sellars et al. (1956) investigated the Sturm-Liouville eigenvalue problem for large eigenvalues for pipe and channel flows with constant wall temperature. They used the WKB(J)[1] method for obtaining the eigenvalues for $\lambda \to \infty$. As a result, they found that the large eigenvalues can be predicted by

$$\lambda_j = \frac{4}{\sqrt{2}}\left(j + \frac{2}{3}\right) \quad ; \quad j = 0, 1, 2, \ldots \tag{4.1}$$

The eigenvalues given by Eq. (4.1) are in good agreement with numerically predicted values for larger j (for $j > 5$ they are nearly identical). Lauwerier (1950) showed, that the solution of the laminar Graetz problem can be expressed by

---

[1] The description WKB(J) stands for <u>W</u>entzel, <u>K</u>ramer, <u>B</u>rillonin and <u>J</u>effreys, who developed the method in the 1920s. However, the method was applied much earlier by Liouville (1837). For an introduction into the method, the reader is referred to Simmonds and Mann (1986) and Plaschko and Brod (1989). For a more detailed description of the method, the reader is referred to Froman and Froman (1965) and to Kumar (1972).

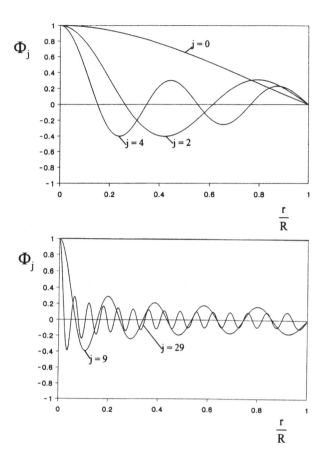

Fig. 4.1: Eigenfunctions for a turbulent pipe flow with constant wall temperature (Re_D = 10000, Pr = 0.01)

Whittaker functions (see Kamke (1983)), where the eigenvalue appears in the argument as well as in the subscript of the function. From the asymptotic behavior of these functions, Lauwerier was able to derive an asymptotic form for the eignfunctions for large eigenvalues. The value of the eigenfunction at the wall was determined to be

$$\Phi(1,\lambda) = \frac{\sqrt{3}}{3\pi}\left(\frac{6}{\sqrt{2\lambda}}\right)^{1/3}\Gamma(1/3)\cos\left[\left(\frac{\sqrt{2\lambda}}{4}-\frac{1}{6}\right)\pi\right]\sum_{p=0}\alpha_p\left(\frac{2}{\sqrt{2\lambda}}\right)^{2p} \qquad (4.2)$$

$$+\frac{\sqrt{3}}{3\pi}\left(\frac{6}{\sqrt{2\lambda}}\right)^{1/3}3^{1/3}\Gamma(2/3)\cos\left[\left(\frac{\sqrt{2\lambda}}{4}+\frac{1}{6}\right)\pi\right]\sum_{p=0}\beta_p\left(\frac{2}{\sqrt{2\lambda}}\right)^{2p+4/3}$$

$$\alpha_0 = 1, \qquad\qquad\qquad \beta_0 = -0.0785714857$$
$$\alpha_1 = -0.01444444444, \qquad \beta_1 = 0.02887167277$$
$$\alpha_2 = 0.009882467268, \qquad \beta_2 = -0.0440553529$$
$$\alpha_3 = -0.02131664753,$$

From Eq. (4.2), the eigenvalues can be calculated for the constant wall temperature case by setting Eq. (4.2) equal to zero. Newman (1969) extended the analysis by Lauwerier and obtained the following expressions

$$\lambda_j = \lambda_0 + s_1 \lambda_0^{-4/3} + s_2 \lambda_0^{-8/3} + s_3 \lambda_0^{-10/3} + s_4 \lambda_0^{-11/3} + O\left(\lambda_0^{-14/3}\right) \tag{4.3}$$

$$A_j \Phi_j'(1) = C \lambda_j^{-1/3} \left[\begin{array}{l} 1 + L_1 \lambda_0^{-4/3} + L_2 \lambda_0^{-6/3} + L_3 \lambda_0^{-7/3} \\ + L_4 \lambda_0^{-10/3} + L_5 \lambda_0^{-11/3} + 0\left(\lambda_0^{-4}\right) \end{array} \right] \tag{4.4}$$

with the quantities

$$\lambda_0 = \frac{4}{\sqrt{2}}\left(j + \frac{2}{3}\right) \quad ; \quad j = 0, 1, 2, \ldots \tag{4.5}$$

$$\begin{array}{ll} s_1 = 0.159152288, & L_1 = 0.144335160 \qquad (4.6) \\ s_2 = 0.0114856354, & L_2 = 0.115555556 \\ s_3 = -0.224731440, & L_3 = -0.21220305 \\ s_4 = -0.033772601, & L_4 = -0.187130142 \\ C = -2.025574576, & L_5 = -0.0918850832 \end{array}$$

For the case of a wall boundary condition of the third kind, Hsu (1971a) derived asymptotic expressions for the eigenvalues and related constants. He showed that his expressions are identical to those of Sellars et al. (1956) for the case of an uniform wall temperature. For the case of an insulated wall, the expressions of Hsu (1971a) are identical to those provided by Hsu (1965) for the case of a constant wall heat flux.

For a turbulent internal flow, the mathematical problem of obtaining asymptotic expressions for the eigenvalues and eigenfunctions is much more complicated than for laminar internal flow. This is due to the complicated description of the velocity profile and the appearance of the eddy diffusivity in Eq. (3.25), which causes the function $a_2(\tilde{n})$ to be a complicated function of the coordinate.

$$\tilde{r}^F \tilde{u}(\tilde{n}) \lambda_j^2 \Phi_j(\tilde{n}) + \left[\tilde{r}^F a_2(\tilde{n}) \Phi_j'(\tilde{n})\right]' = 0 \tag{3.25}$$

$$\tilde{n} = 0 : \Phi'(0) = 0$$
$$\tilde{n} = 1 : \Phi(1) = 0$$
(3.26)

Sternling and Sleicher (1962) derived asymptotic expressions for the eigenvalues and eigenfunctions for heat transfer in a turbulent pipe flow with constant wall temperature. However, it turned out that the accuracy of their results for A_j, Eq. (3.59), as well as for $A_j\Phi_j(1)$, Eq. (3.69), was somewhat disappointing. Eight years later, Sleicher et al. (1970) were able to derive more accurate expressions using a matched asymptotic expansion. In the following sections, the method of Sleicher et al. (1970) is described in more detail for the case of a constant wall temperature, because their method is very useful for a lot of similar problems. The method has been successfully used with some modifications by Shibani and Özisik (1977a), Sadeghipour et al. (1984) and Shibani and Özisik (1977b) for different wall boundary conditions and different geometries. At the time - the method has been established by Sleicher et al. (1970) - the computational power of the computers did not permit a detailed comparison between their derived asymptotic expressions and numerical solutions of the eigenvalue problem for larger eigenvalues. Therefore, a short section has been added at the end of this chapter which shows a detailed evaluation of the asymptotic expressions compared to numerical calculations. In addition, it is shown, that the method of Sleicher et al. (1970) can be adapted quite easily to other related problems. This is demonstrated for the heat transfer process in an axially rotating pipe.

We start our considerations with the Sturm-Liouville problem given by the Eqs. (3.25-3.26) and consider a turbulent pipe flow. For this type of application, the equations read ($F = 1$, $\tilde{n} = \tilde{r}$)

$$\frac{d}{d\tilde{r}}\left[\tilde{r}\,a_2(\tilde{r})\frac{d\Phi_j}{d\tilde{r}}\right] + \tilde{r}\,\tilde{u}(\tilde{r})\lambda_j^2\Phi_j(\tilde{r}) = 0 \tag{4.7}$$

with the boundary conditions

$$\tilde{r} = 0 : \Phi'(0) = 0 \tag{4.8}$$
$$\tilde{r} = 1 : \Phi(1) = 0$$

In order to obtain an asymptotic expression for large eigenvalues, it is common to transform the Sturm-Liouville system into its standard form. This can be done by introducing the following new quantities (see Kamke (1983))

$$G = \frac{1}{\pi}\int_0^1\sqrt{\frac{\tilde{u}}{2a_2}}\,d\tilde{r}, \quad k = \sqrt{2}\lambda_j G, \quad \xi = \frac{1}{G}\int_0^{\tilde{r}}\sqrt{\frac{\tilde{u}}{2a_2}}\,d\tilde{r} \tag{4.9}$$

$$\vartheta = \left(\frac{\tilde{r}^2 a_2\tilde{u}}{2}\right)^{1/4}\Phi_j, \quad w(\xi) = \left(\frac{\tilde{r}^2 a_2\tilde{u}}{2}\right)^{-1/4}\frac{d^2}{d\xi^2}\left(\frac{\tilde{r}^2 a_2\tilde{u}}{2}\right)^{1/4} \tag{4.10}$$

One obtains

$$\frac{d^2 \vartheta}{d\xi^2} + \left[k^2 - w(\xi)\right]\vartheta = 0 \tag{4.11}$$

with the boundary conditions

$$\xi = 0 : \vartheta(0) = 0 \tag{4.12}$$
$$\xi = \pi : \vartheta(\pi) = 0$$

After introducing the new coordinate ξ, the interval for the coordinate has changed from [0,1] to [0,π]. Before trying to find a solution for the eigenfunction ϑ for larger eigenvalues, one has to analyze the function $w(\xi)$ in greater detail: the function $w(\xi)$ has singularities at both boundaries $(\xi = 0, \xi = \pi)$. However, the function can be approximated near both boundaries with sufficient accuracy. Sternling and Sleicher (1962) subdivided the interval into two regions (core region and near wall region). For both regions, they determined solutions for Eq. (4.11) and patched them together. This procedure resulted, however, in a disappointing accuracy for the constants A_j. Therefore, Sleicher et al. (1970) choose a different approach. The interval for ξ [0,π] is split up in three separate regions (core region, near wall region and middle region). The solutions for the separate regions have to be connected later to each other by expansions of the variables[2].

In order to find a solution for Eqs. (4.11-4.12) for large eigenvalues, one can use the following expansion (see van Dyke (1962) or Kevorkian and Cole (1981))

$$\vartheta \sim \sum_{i=0}^{N} \delta_i(k) \, \vartheta_i(\xi) \tag{4.13}$$

The functions $\delta_i(k)$ are a priori not known and have to be determined during the analysis. In order to show more clearly the solution procedure of Sleicher et al. (1970), these functions are introduced here from the very beginning. Later, during the calculation procedure, we clarify why exactly this sequence of functions is chosen for the present example

$$\delta_0 = k^{-1/2}, \quad \delta_1 = k^{-3/2}\ln k, \quad \delta_2 = k^{-3/2}, \quad \delta_3 = k^{-5/2}\ln k \tag{4.14}$$
$$\delta_4 = k^{-5/2}, \quad \delta_5 = k^{-5/2}\left(\ln k\right)^2$$

Introducing Eq. (4.14) into Eq. (4.13) one obtains the following expression for the eigenfunctions

$$\vartheta \sim k^{-1/2}\vartheta_0 + k^{-3/2}\ln k\,\vartheta_1 + k^{-3/2}\vartheta_2 + \tag{4.15}$$
$$+ k^{-5/2}\ln k\,\vartheta_3 + k^{-5/2}\vartheta_4 + k^{-5/2}\left(\ln k\right)^2\vartheta_5$$

---

[2] The analysis presented here follows the excellent work of Sleicher et al. (1970). It should serve as an example on how to solve such an difficult eigenvalue propblem.

After describing the asymptotic expansion coefficients for ϑ, approximations for the function $w(\xi)$ have to be derived. Because of the singularities of the function near the boundaries, the behavior of $w(\xi)$ has to be examined separately for all three regions.

Pipe Center Region

Near the center of the pipe, the axial velocity component can be approximated with good accuracy by a constant value U. The distribution for a_2 can be approximated for this area by a constant A. If one introduces this assumptions into Eqs. (4.9-4.10) one obtains:

$$\xi = \pi \tilde{r} \tag{4.16}$$

$$w(\xi) = \xi^{-1/2} \frac{d^2}{d\xi^2} (\xi)^{1/2} = -\frac{1}{4}\xi^{-2} \tag{4.17}$$

Middle Region

In this area, the function $w(\xi)$ has no singularities. Since we are interested in solutions for large eigenvalues, the function w can be neglected compared to k^2 in Eq. (4.11).

Near Wall Region

At the boundary ($\xi = \pi$), the function $w(\xi)$ has a second singularity. Very close to the wall, the velocity can be approximated by $\tilde{u} \sim \tilde{y}_W = (1 - \tilde{r})$. The function a_2 tends to one in this region. If we introduce the above assumptions into Eq. (4.10), we obtain for $w(\xi)$ after neglecting quadratic terms

$$w(\xi) \approx -\frac{\beta_1}{(\pi - \xi)^2}, \quad \beta_1 = \frac{5}{36} \tag{4.18}$$

The above expression for $w(\xi)$ was first derived by Sternling and Sleicher (1962)[3]. Clearly, the region, where Eq. (4.18) is fully valid, is very small. There-

---

[3] The constant β_1 is determined by the velocity distribution near the wall. The value of the constant is obtained under the assumption of a linear velocity distribution of the form

$$\tilde{u} = 2V \tilde{y}_W, \quad V = \frac{1}{2}\left(\frac{d\tilde{u}}{d\tilde{y}_W}\right)_{\tilde{y}_W = 0} = \frac{\text{Re}_D \, c_f}{32}$$

fore, Sleicher et al. (1970) used the following expression for approximating the function $w(\xi)$ in the near wall region

$$w(\xi) \approx -\frac{\beta_1}{(\pi - \xi)^2} + \frac{\beta_2}{\pi - \xi} \tag{4.19}$$

Equation (4.19) assumes that the function $w(\xi)$ can be expanded into a Laurent series around $\xi = \pi$. Equation (4.19) contains the not yet determined constant β_2. This constant is determined during the solution process.

After having investigated the functional form of $w(\xi)$ for the different regions, the determination of the eigenfunctions from Eq. (4.11) for the three different regions can be started.

Pipe Center Region

In the pipe center, the function $w(\xi)$ is given by Eq. (4.17). Inserting this expression into Eq. (4.11) results in

$$\frac{d^2\vartheta_C}{ds^2} + \left[1 + \frac{1}{4s^2}\right]\vartheta_C = 0 \tag{4.20}$$

where the stretched coordinate

$$s = k\xi \tag{4.21}$$

is used. The new coordinate s has the advantage that the eigenvalue does not appear explicitly in the differential equation. Figure 4.2 shows the different coordinates and the geometry under consideration.

Introducing the expansion defined in Eq. (4.15) into Eq. (4.20) results after collecting terms of the same order in k in the following set of equations

$$\frac{d^2\vartheta_{iC}}{ds^2} + \left[1 + \frac{1}{4s^2}\right]\vartheta_{iC} = 0, \quad i = 0, 1, ..., 5 \tag{4.22}$$

---

where c_f is a friction factor (Sleicher et al. (1970)). If one uses a velocity distribution according to

$$\tilde{u} = 2c(1 - \tilde{r})^{1/m}, \quad c = \frac{1}{4}\left(\frac{3m + 1}{m^2} + 2\right)$$

one obtains for the constant β_1

$$\beta_1 = \frac{1}{4}\frac{1 + 4m}{(1 + 2m)^2}$$

The reader is referred to Shibani and Özisik (1977a) for this approach.

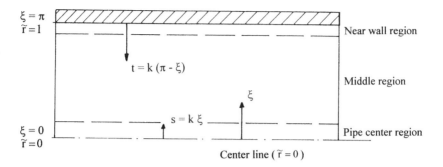

Fig. 4.2: Geometry and different coordinate systems used

Equation (4.22) can be transformed into a Bessel equation after introducing the new function

$$\vartheta_{iC} = \sqrt{s}\ \tilde{\vartheta}_{iC} \tag{4.23}$$

The solution of this Bessel equation, which is finite for $s = 0$, is given by

$$\vartheta_{iC} = B_i \sqrt{s}\ J_0(s), \quad i - 0,1,...,5 \tag{4.24}$$

The complete solution for the pipe center region is therefore given by

$$\vartheta_C = \sqrt{s}\ J_0(s) \left\{ \begin{array}{l} B_0\ k^{-1/2} + B_1\ k^{-3/2}\ \ln k + B_2\ k^{-3/2} + \\ B_3\ k^{-5/2}\ \ln k + B_4\ k^{-5/2} + B_5\ k^{-5/2}\left(\ln k\right)^2 \end{array} \right\} \tag{4.25}$$

From the definition of the function ϑ in Eq. (4.10) it follows that in the pipe center ($\tilde{r} \to 0$, $a_2 = A$, $\tilde{u} = U$, $\Phi_j(0) = 1$), the function ϑ will behave like

$$\xi \to 0 : \vartheta_C = \left(\frac{\tilde{r}^2 a_2 \tilde{u}}{2}\right)^{1/4} \Phi_j(0) = \sqrt{GA\xi} \tag{4.26}$$

where the normalizing condition for the eigenfunction $\Phi_j(0) = 1$ has been used. If one introduces now the stretched coordinate s into Eq. (4.26), one obtains in the center of the pipe

$$\xi \to 0 : \vartheta_C(0) = \sqrt{GA\xi} \Rightarrow, \quad s \to 0 : k^{-1/2}\sqrt{GA}\ \sqrt{s} \tag{4.27}$$

This expression can be compared to Eq. (4.25) for $s \to 0$ ($J_0(0) = 1$). This results in

$$B_0 + B_1 k^{-1} \ln k + B_2 k^{-1} + B_3 k^{-2} \ln k + B_4 k^{-2} + B_5 k^{-2} \left(\ln k \right)^2 = \sqrt{GA} \qquad (4.28)$$

Because Eq. (4.28) has to be valid for all k, it follows

$$B_0 = \sqrt{GA} \qquad (4.29)$$
$$B_i = 0, \quad i = 1, ..., 5$$

And the solution for the center region of the pipe is

$$\vartheta_C = k^{-1/2} \sqrt{GA} \sqrt{s} \, J_0 \left(s \right) \qquad (4.30)$$

Middle Region

In the middle region, the function $w(\xi)$ is a continuous function without singularities. Because of this, $w(\xi)$ can be neglected in Eq. (4.11) compared to the large eigenvalues. After introducing again the stretched coordinate s, one obtains from Eq. (4.11)

$$\frac{d^2 \vartheta_M}{ds^2} + \vartheta_M = 0 \qquad (4.31)$$

Introducing the expansion according to Eq. (4.15) into Eq. (4.31) results after collecting terms of the same order in k, in the following set of equations

$$\frac{d^2 \vartheta_{iM}}{ds^2} + \vartheta_{iM} = 0, \quad i = 0, 1, ..., 5 \qquad (4.32)$$

The solution of Eq.(4.32) can be calculated easily and is given by

$$\vartheta_M = k^{-1/2} \left(K_0 \sin s + L_0 \cos s \right) + k^{-3/2} \ln k \left(K_1 \sin s + L_1 \cos s \right) + \qquad (4.33)$$
$$k^{-3/2} \left(K_2 \sin s + L_2 \cos s \right) + k^{-5/2} \ln k \left(K_3 \sin s + L_3 \cos s \right) +$$
$$k^{-5/2} \left(K_4 \sin s + L_4 \cos s \right) + k^{-5/2} \left(\ln k \right)^2 \left(K_5 \sin s + L_5 \cos s \right)$$

The constants K_i and L_i, which appear in Eq. (4.33), are yet not known. In contrast to the constants in the solution for the pipe center, they have to be determined during the following solution process.

Near Wall Region

For the near wall region, the function $w(\xi)$ is given by Eq. (4.19). Inserting this expression into Eq. (4.11) results in

$$\frac{d^2 \vartheta_W}{dt^2} + \left[1 + \frac{\beta_1}{t^2} - \frac{\beta_2}{kt}\right]\vartheta_W = 0 \tag{4.34}$$

where the new stretched coordinate

$$t = k\left(\pi - \xi\right) = k\pi - s \tag{4.35}$$

is introduced. Inserting the expansion according to Eq. (4.15) into Eq. (4.35) results, after collecting terms of the same order in k, in the following set of equations

$$\frac{d^2 \vartheta_{iW}}{dt^2} + \left[1 + \frac{\beta_1}{t^2}\right]\vartheta_{iW} = 0, \quad i = 0,1 \tag{4.36}$$

$$\frac{d^2 \vartheta_{2W}}{dt^2} + \left[1 + \frac{\beta_1}{t^2}\right]\vartheta_{2W} = \frac{\beta_2}{t} \vartheta_{0W} \tag{4.37}$$

$$\frac{d^2 \vartheta_{3W}}{dt^2} + \left[1 + \frac{\beta_1}{t^2}\right]\vartheta_{3W} = \frac{\beta_2}{t} \vartheta_{1W} \tag{4.38}$$

$$\frac{d^2 \vartheta_{4W}}{dt^2} + \left[1 + \frac{\beta_1}{t^2}\right]\vartheta_{4W} = \frac{\beta_2}{t} \vartheta_{2W} \tag{4.39}$$

$$\frac{d^2 \vartheta_{5W}}{dt^2} + \left[1 + \frac{\beta_1}{t^2}\right]\vartheta_{5W} = 0 \tag{4.40}$$

The solution of the homogenous differential equations (4.36) and (4.40) can be obtained in the same way as for Eq. (4.22). Introducing again the new function

$$\vartheta_{iW} = \sqrt{t}\, \tilde{\vartheta}_{iW} \tag{4.41}$$

into Eqs. (4.36) and (4.40) results in

$$t^2 \vartheta_{iW}'' + t\vartheta_{iW}' + \left(t^2 + \left(\beta_1 - 1/4\right)\right)\vartheta_{iW} = 0 \tag{4.42}$$

As it can be seen, Eq. (4.42) is a Bessel equation (see Bowman (1958)). For $\beta_1 = 5/36$ [4] (this is the value used by Sternling and Sleicher (1962) for the linear velocity distribution near the wall) one obtains

$$t^2 \vartheta_{iW}'' + t\vartheta_{iW}' + \left(t^2 + 1/9\right)\vartheta_{iW} = 0 \tag{4.43}$$

This equation has the general solution

$$\vartheta_{iW} = \sqrt{t}\left\{D_i\, J_{1/3}\left(t\right) + E_i\, J_{-1/3}\left(t\right)\right\}, \qquad i = 0,1,5 \tag{4.44}$$

---

[4] For other values of β_1, the reader can find some solutions in Shibani and Özisik (1977a).

Equation (4.44) is also the solution of the homogenous Eqs. (4.37)-(4.39). A particular solution for these equations can be constructed by using the method of the variation of constants (see Bronstein and Semendjajew (1981)). This leads finally to the complete solution for the near wall region, given by

$$
\vartheta_w = k^{-1/2} \sqrt{t} \left\{ D_0 \, J_{1/3}(t) + E_0 \, J_{-1/3}(t) \right\} \tag{4.45}
$$
$$
+ k^{-3/2} \ln k \sqrt{t} \left\{ D_1 \, J_{1/3}(t) + E_1 \, J_{-1/3}(t) \right\} +
$$
$$
k^{-3/2} \sqrt{t} \, J_{1/3} \left\{ \frac{\pi}{\sqrt{3}} \beta_2 \int \left(\begin{array}{c} D_0 \, J_{1/3} \, J_{-1/3} + \\ E_0 \, J_{-1/3} \, J_{-1/3} \end{array} \right) d\tilde{t} + D_2 \right\} +
$$
$$
- k^{-3/2} \sqrt{t} \, J_{-1/3} \left\{ \frac{\pi}{\sqrt{3}} \beta_2 \int \left(\begin{array}{c} D_0 \, J_{1/3} \, J_{1/3} \\ + E_0 \, J_{1/3} \, J_{-1/3} \end{array} \right) d\tilde{t} + E_2 \right\} +
$$
$$
k^{-5/2} \ln k \sqrt{t} \, J_{1/3} \left\{ \frac{\pi}{\sqrt{3}} \beta_2 \int \left(\begin{array}{c} D_1 \, J_{1/3} \, J_{-1/3} \\ + E_1 \, J_{-1/3} \, J_{-1/3} \end{array} \right) d\tilde{t} + D_3 \right\}
$$
$$
- k^{-5/2} \ln k \sqrt{t} \, J_{-1/3} \left\{ \frac{\pi}{\sqrt{3}} \beta_2 \int \left(\begin{array}{c} D_1 \, J_{1/3} \, J_{1/3} \\ + E_1 \, J_{-1/3} \, J_{1/3} \end{array} \right) d\tilde{t} + E_3 \right\}
$$
$$
+ k^{-5/2} \frac{\pi \beta_2 \sqrt{t}}{\sqrt{3}} J_{1/3} \cdot
$$
$$
\left[\begin{array}{c} \int J_{1/3} \, J_{-1/3} \left\{ \frac{\pi}{\sqrt{3}} \int \beta_2 \left(\begin{array}{c} D_0 \, J_{1/3} \, J_{-1/3} \\ + E_0 \, J_{-1/3} \, J_{-1/3} \end{array} \right) d\tilde{t} + D_2 \right\} d\tilde{\tilde{t}} \\ - \int J_{-1/3} \, J_{-1/3} \left\{ \frac{\pi}{\sqrt{3}} \int \beta_2 \left(\begin{array}{c} D_0 \, J_{1/3} \, J_{1/3} \\ + E_0 \, J_{-1/3} \, J_{1/3} \end{array} \right) d\tilde{t} + E_2 \right\} d\tilde{\tilde{t}} + D_4 \end{array} \right]
$$
$$
- k^{-5/2} \frac{\pi \, \beta_2 \sqrt{t}}{\sqrt{3}} J_{-1/3} \cdot
$$
$$
\left[\begin{array}{c} \int J_{1/3} \, J_{1/3} \left\{ \frac{\pi}{\sqrt{3}} \int \beta_2 \left(\begin{array}{c} D_0 \, J_{1/3} \, J_{-1/3} \\ + E_0 \, J_{-1/3} \, J_{-1/3} \end{array} \right) d\tilde{t} + D_2 \right\} d\tilde{\tilde{t}} - \\ \int J_{-1/3} \, J_{1/3} \left\{ \frac{\pi}{\sqrt{3}} \int \beta_2 \left(\begin{array}{c} D_0 \, J_{1/3} \, J_{1/3} \\ + E_0 \, J_{-1/3} \, J_{1/3} \end{array} \right) d\tilde{t} + E_2 \right\} d\tilde{\tilde{t}} + E_4 \end{array} \right]
$$
$$
+ k^{-5/2} (\ln k)^2 \sqrt{t} \left\{ D_5 \, J_{1/3}(t) + E_5 \, J_{-1/3}(t) \right\}
$$

The constants D_i and E_i, appearing in Eq. (4.45) have to be determined during the following matching procedure for the individual solutions in the different regions.

The matching of the individual solutions is done for overlapping regions, where both solutions are still valid. Here, one increases the eigenvalue for a fixed value of the coordinate. The matching of the solution is demonstrated for simplicity only for the first three term of the expansion

$$\vartheta \sim k^{-1/2}\vartheta_0 + k^{-3/2}\ln k \,\vartheta_1 + k^{-3/2}\vartheta_2 \qquad (4.46)$$

This has the advantage that the basic ideas behind the matching procedure of the solutions can be shown clearly. The algebra involved by using the expression given by Eq. (4.15) is much more complicated and does not contribute to a better understanding of the method (see also Sleicher et al. (1970)).

From the previously obtained solution for ϑ_M, Eq. (4.33), it is clear that trigonometric functions have to be matched within the solution process. Because of this fact, it is important to investigate the approximation of these functions in greater detail. As it is already known, trigonometric functions have an amplitude, a frequency and a phase shift. The question now is whether, in approximating these functions, a small error in amplitude or phase shift (for a given frequency) is allowed. Fig. 4.3 shows a function g and two different approximations for this function (g_1 and g_2). For g_1, a small deviation in amplitude is allowed compared to g, whereas for g_2, a small deviation in phase shift is allowed compared to g.

It can be seen clearly that the function g_1 represents a good approximation of g, even though the amplitude of g_1 is slightly wrong. On the other side, the function g_2 does not approximate g very well, because the difference in the phase shift leads to large errors. This shows clearly that for the matching procedure the phase shift has to be matched always exactly. Small deviations in the amplitude can be allowed, because they lead to satisfactory approximations of the functions. Because we found already a complete solution for ϑ_C, it is logical to start the process with the two functions ϑ_C and ϑ_M.

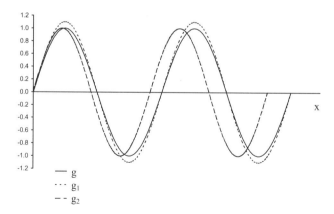

Fig. 4.3: Influence of a phase shift and a changed amplitude for a trigonometric function

If we investigate Eq. (4.30) for very large values of the eigenvalue, the following asymptotic expansion for the Bessel function can be used (Gradshteyn and Ryzhik (1994))

$$
J_{\pm i}(s) = \sqrt{\frac{2}{\pi s}} \left\{ \begin{array}{l} \cos\left(s \mp \frac{\pi}{2}i - \frac{\pi}{4}\right) \left[\sum_{n=0}^{M-1} \frac{(-1)^n}{(2s)^{2n}} \frac{\Gamma(i+2n+1/2)}{(2n)!\,\Gamma(i-2n+1/2)} + R_1 \right] - \\ \sin\left(s \mp \frac{\pi}{2}i - \frac{\pi}{4}\right) \left[\sum_{n=0}^{M-1} \frac{(-1)^k}{(2s)^{k+1}} \frac{\Gamma(i+2n+3/2)}{(2n+1)!\,\Gamma(i-2n-1/2)} + R_2 \right] \end{array} \right\}
\tag{4.47}
$$

where the two quantities R_1 and R_2 denote the truncation errors of the sums in Eq. (4.47). These errors are of the order of M. Using Eq. (4.47), the distribution in the center region of the pipe, given by Eq. (4.30), can be approximated for large eigenvalues by

$$
\vartheta_C = k^{-1/2} \sqrt{\frac{2GA}{\pi}} \left\{ \cos(s - \pi/4) + \frac{1}{8s}\sin(s - \pi/4) \right\}
\tag{4.48}
$$

If one now equates Eq. (4.48) and Eq. (4.33) (which are the solutions for the center region and the middle region), one obtains an equation for determining the unknown constants K_i and L_i

$$
\sqrt{\frac{2GA}{\pi}} \left\{ \cos(s - \pi/4) + \underline{\frac{1}{8s}\sin(s - \pi/4)} \right\} = K_0 \sin s + L_0 \cos s +
\tag{4.49}
$$

$$
\frac{\ln k}{k}\{K_1 \sin s + L_1 \cos s\} + \frac{1}{k}\{K_2 \sin s + L_2 \cos s\}
$$

As stated before, only terms up to the order $O\left(k^{-3/2}\right)$ are considered here for simplicity. As it can be seen from Eq. (4.49), the coordinate s appears only for the underlined term in the amplitude. The stretched coordinate s, rewritten as a function of the second stretched coordinate t, according to Eq. (4.35), is $s = k\pi - t$. Matching of the two solutions ϑ_C and ϑ_M has now to be done for a fixed value of t. This means that s tends to $k\pi$ for large k. This means also that the amplitude $1/(8s)$ in Eq. (4.49) can be approximated by $1/(8k\pi)$. Applying the addition theorem for trigonometric functions, one obtains from Eq. (4.49)

$$
\sqrt{\frac{GA}{\pi}} \left\{ \cos s + \sin s + \frac{1}{k}\frac{1}{8\pi}[\sin s - \cos s\,] \right\} = K_0 \sin s
\tag{4.50}
$$

$$
+ L_0 \cos s + \frac{\ln k}{k}\{K_1 \sin s + L_1 \cos s\} + \frac{1}{k}\{K_2 \sin s + L_2 \cos s\}
$$

Comparing the terms in Eq. (4.50) which are of order 1 results in

$$K_0 = L_0 = \sqrt{\frac{GA}{\pi}} \tag{4.51}$$

Comparing terms in Eq. (4.50), which are of the order $1/k \ln k$ and $1/k$ leads to

$$K_1 = L_1 = 0 \tag{4.52}$$

$$K_2 = -L_2 = \frac{1}{8\pi} \sqrt{\frac{GA}{\pi}} \tag{4.53}$$

After the constants K_i and L_i have been determined, the solution for the middle region can be matched to the solution for the near wall region. This gives us the unknown coefficients D_i and E_i. In order to do this, the Bessel functions, appearing in Eq. (4.45), have to be approximated for large eigenvalues by the series expansion, Eq. (4.47). This leads, after equating the solutions for the near wall region and the middle region, to the following expression

$$
\begin{aligned}
&D_0 \left\{ \cos\left(t - 5\pi/12\right) + \frac{5}{72t} \sin\left(t - 5\pi/12\right) \right\} + \\
&E_0 \left\{ \cos\left(t - \pi/12\right) + \frac{5}{72t} \sin\left(t - \pi/12\right) \right\} \\
&\frac{\ln k}{k} \left\{ D_1 \cos\left(t - 5\pi/12\right) + E_1 \sin\left(t - 5\pi/12\right) \right\} + \\
&\frac{1}{k} \left\{ \cos\left(t - 5\pi/12\right) \left[D_2 - \left(\frac{\beta_2 D_0}{\sqrt{3}} + \frac{\beta_2 E_0}{\sqrt{3}} \right) \ln t \right] \right\} - \\
&\frac{1}{k} \left\{ \cos\left(t - \pi/12\right) \left[E_2 - \left(\frac{\beta_2 D_0}{\sqrt{3}} + \frac{\beta_2 E_0}{\sqrt{3}} \right) \ln t \right] \right\} = \\
&\sqrt{\frac{\pi}{2}} \left[K_0 \sin s + L_0 \cos s + \frac{1}{k} \left(K_2 \sin s + L_2 \cos s \right) \right]
\end{aligned}
\tag{4.54}
$$

From Eq. (4.54), it is immediately clear why the expansion given in Eq. (4.15) contained also logarithmic terms. This terms are needed, because in the asymptotic expansion of the solution for the near wall region, ϑ_W, such terms are present if $w(\xi)$ is approximated by Eq. (4.19). Now the matching procedure works exactly as explained for Eq. (4.49). The relation between the two stretched coordinates s and t is given by Eq. (4.35). The two solutions ϑ_W and ϑ_M have to be matched for a fixed value of s and for $k \to \infty$. This means that in the amplitude t can be replaced by $k\pi$ and $\ln(t)$ by $\ln(k)$. However, as pointed out before, these simplifications are only possible for the amplitude of the function. It must not be done for the phase shift. After selecting terms of order k^0 in Eq.(4.54), we obtain

$$D_0 \cos\left(t - 5\pi/12\right) + E_0 \cos\left(t - 5\pi/12\right) = \tag{4.55}$$

$$\frac{\sqrt{2}}{\pi}\left[K_0 \sin\left(k\pi - t\right) + L_0 \cos\left(k\pi - t\right)\right]$$

Using the addition theorem for the trigonometric functions, one obtains the two unknown constants D_0 and E_0

$$D_0 = \left(\frac{4GA}{3}\right)^{1/2} \sin\left(k\pi - \pi/3\right), \tag{4.56}$$

$$E_0 = -\left(\frac{4GA}{3}\right)^{1/2} \sin\left(k\pi - 2\pi/3\right)$$

Selecting terms of the order $k^{-1} \ln k$ in Eq. (4.54) leads to

$$D_1 = \beta_2 \left(\frac{GA}{3}\right)^{1/2} \cos\left(k\pi - \pi/3\right), \tag{4.57}$$

$$E_1 = \beta_2 \left(\frac{GA}{3}\right)^{1/2} \cos\left(k\pi - 2\pi/3\right)$$

Finally, selecting terms of the order k^{-1} in Eq. (4.54) results in

$$D_2 = -\left(\frac{GA}{3}\right)^{1/2} \cos\left(k\pi - \pi/3\right)\left(\frac{7}{18\pi} + \beta_2 \ln\pi\right), \tag{4.58}$$

$$E_2 = -\left(\frac{GA}{3}\right)^{1/2} \cos\left(k\pi - 2\pi/3\right)\left(\frac{7}{18\pi} + \beta_2 \ln\pi\right)$$

After this, all the unknown constants have been obtained. If the six term expansion is used instead of the three term expansion, the remaining constants K_i, L_i, D_i, E_i, $i=3,4,5$ can be determined in the same way. For this constants one obtains (see Sleicher et al. (1970))

$$K_3 = L_3 = 0, \quad K_4 = L_4 = \frac{1}{\pi^2}\sqrt{\frac{GA}{\pi}}, \quad K_5 = L_5 = 0 \tag{4.59}$$

$$D_3 = -\frac{\beta_2}{2}\left(\frac{GA}{3}\right)^{1/2} \sin\left(k\pi - \pi/3\right)\left(\frac{7}{18\pi} + \beta_2 \ln\pi\right), \tag{4.60}$$

$$E_3 = -\frac{\beta_2}{2}\left(\frac{GA}{3}\right)^{1/2} \sin\left(k\pi - 2\pi/3\right)\left(\frac{7}{18\pi} + \beta_2 \ln\pi\right)$$

$$D_4 = -\left(\frac{GA}{3}\right)^{1/2} \sin\left(k\pi - \pi/3\right)\left(\frac{\beta_2}{2\pi} + \frac{121}{1296\pi^2} + \frac{7\beta_2 \ln\pi}{36\pi} + \frac{\left(\beta_2 \ln\pi\right)^2}{4}\right), \quad (4.61)$$

$$E_4 = -\left(\frac{GA}{3}\right)^{1/2} \sin\left(k\pi - 2\pi/3\right)\left(\frac{\beta_2}{2\pi} + \frac{121}{1296\pi^2} + \frac{7\beta_2 \ln\pi}{36\pi} + \frac{\left(\beta_2 \ln\pi\right)^2}{4}\right)$$

$$D_5 = \frac{\beta_2^2}{4}\left(\frac{GA}{3}\right)^{1/2} \sin\left(k\pi - \pi/3\right), \quad (4.62)$$

$$E_5 = \frac{\beta_2^2}{4}\left(\frac{GA}{3}\right)^{1/2} \sin\left(k\pi - 2\pi/3\right)$$

After the eigenfunctions are completely known, the eigenvalues can be calculated by setting the eigenfunction, Eq. (4.45), at the pipe wall equal to zero. One obtains

$$E_0 + \frac{\ln k}{k} E_1 - \frac{1}{k} E_2 = 0 \quad (4.63)$$

For evaluating the eigenvalues only three terms of the series have been considered here, because this guarantees a satisfactory accuracy (Sleicher et al. (1970)). If one introduces the expressions for the constants E_i, one finally obtains the following equations for determining the eigenvalues

$$2\tan\left(k\pi - \frac{2\pi}{3}\right) - \frac{\ln k}{k}\beta_2 + \frac{1}{k}\left(\frac{7}{18\pi} + \beta_2 \ln\pi\right) = 0 \quad (4.64)$$

Eq. (4.64) is an implicit equation for determining the eigenvalues. Sleicher et al. (1970) used several approximations to derive an explicit equation. As a final result they obtained

$$k_j = j + \frac{2}{3} + \frac{\beta_2 \ln\left(j\pi + 2\pi/3\right) + 7/\left(18\pi\right)}{2\pi\left(j + 2/3\right)} \quad (4.65)$$

The constants A_j, which appear in the temperature distribution Eq. (3.55), can be evaluated from the integrals of the eigenfunctions given by Eq. (3.57). Alternatively, one can derive from the orthogonality of the eigenfunctions the following equation for the constants A_j

$$A_j = \frac{-2}{\lambda_j} / \left.\frac{d\Phi_j}{d\lambda}\right|_{\lambda_j} \quad (4.66)$$

If Eq. (4.66) is applied to the solution for the near wall region, Eq. (4.45), one obtains

$$A_j = \frac{(-1)^j \Gamma(2/3) 3^{2/3} V^{1/6}}{G^{1/3} \pi \sqrt{2GA} \, k_j^{2/3} 2^{1/6}} \bigg/ \left[1 + \frac{1}{k_j^2} \left(\frac{\beta_2}{2\pi} \left(\ln k_j \pi - 3/2 \right) + \frac{1}{6\pi^2} \right) \right]$$

(4.67)

For the determination of the constants A_j, the complete six term expansion has been used. For technical applications, the engineer may be interested only in the distribution of the Nusselt number. As it can be seen from Eq. (3.69), only the product $A_j \Phi'_j(1) = M_j$ appears in this equation. The quantity M_j can be predicted using the six term expansion

$$M_j = \frac{-\Gamma(2/3) V^{1/3}}{G^{2/3} \pi \, k_j^{1/3} 3^{1/6} \Gamma(4/3)} \left[1 - \frac{1}{k_j^2} \left(\frac{\beta_2}{2\pi} \left(\ln k_j \pi - 1 \right) + \frac{7}{36\pi^2} \right) \right]$$

(4.68)

As it can be seen from the Eqs. (4.67-4.68), the unknown constant β_2 is still part of the solution. Sleicher et al. (1970) proposed that the constant β_2 should be evaluated in such a way that the numerically predicted eigenvalues for $j = 3$ are identical to the asymptotic expressions given above (see Notter and Sleicher (1972) for tabulated values of β_2 for different Reynolds and Prandtl numbers). This procedure has the advantage that one receives a continuous spectrum of eigenvalues and constants. However, there is the disadvantage, that the asymptotic behavior is enforced even though it may not be correct at such low values of j. On the other hand, Eqs. (4.67-4.68) show that the term containing the constant β_2 decays at least like $k^{-2} \ln k\pi$. This means that the expressions for the constants are always correct for very large values of j, independently from which value of β_2 has been used.

In the following section, the asymptotic values are compared to numerical calculated values for the Sturm-Liouville system. This is done first for the case of a circular pipe with constant wall temperature and hydrodynamically fully developed turbulent internal flow. The second case shows an application of the method to another problem (heat transfer in an axially rotating pipe).

4.1 Heat Transfer in Turbulent Pipe Flow with Constant Wall Temperature

As stated before, the here considered case of a hydrodynamically fully developed-turbulent pipe flow subjected to constant wall temperature is exactly the one for which Sleicher et al. (1970) developed their asymptotic method. Before comparing the values from the asymptotic expansion to our numerical calculations, it is important to understand first for which combinations of Re_D and Pr a large number of eigenvalues is needed for predicting the distribution of the Nusselt number correctly. Fig. 4.4 shows the influence of the Prandtl number on the length of the thermal entrance for a turbulent flow in a duct and for a given value of the Reynolds number. It can be seen that with increasing values of the Prandtl number the length of the thermal entrance region decreases dramatically.

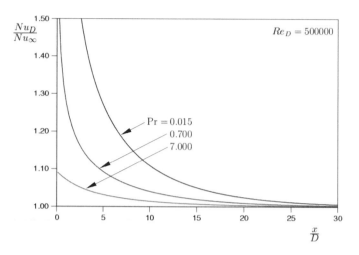

Fig. 4.4: Influence of the Prandtl number on the distribution of the Nusselt number in the thermal entrance region of a turbulent flow in a duct

This shows nicely that for high Prandtl numbers, the value of the fully developed Nusselt number Nu_∞ is sufficient to describe the heat transfer problem. On the other hand, it can be seen that for fluids with relatively low Prandtl numbers (for example for liquid metals ($Pr < 0.1$)) a lot of eigenvalues are necessary for correctly describing the distribution of the Nusselt number in the thermal entrance region. Because of this fact, the focal point of the following study is set on flows with low Prandtl numbers. Sleicher et al. (1970) did not show, for the case of a constant wall temperature, how the asymptotic expressions for λ_j, A_j and M_j compare to numerical predictions of the problem. However, such a comparison is very useful in order to gain confidence in the method shown before. For the first calculations, the value of β_2 is set to zero in the Eqs. (4.65, 4.67-4.68). For the calculation of the eigenvalues only the simple expression $k_j = j + 2/3$ has been used. The case $\beta_2 = 0$ presents the maximum deviation between numerical results and the analytical solution and is, therefore, very important. In order to minimize the error in the numerical calculations, all calculations have been performed with very fine grids (1000-1500 grid points in the radial direction, see Appendix C for a description of the numerical method). In addition, the constants A_j have been calculated separately from Eq. (3.57) and from (4.65). Table 4.1 and Table 4.2 show comparisons of the eigenvalues $\lambda_5, \lambda_{10}, \lambda_{20}$ and the constants M_5, M_{10}, M_{20} for different Prandtl numbers. It can be seen that the eigenvalues can be predicted quite accurately for all Prandtl numbers by using the simple approximation $k_j = j + 2/3$. The maximum deviation between numerically calculated eigenvalues and this simple expression is about 3%. For the constants M_j, which are used in calculating the Nusselt number, the behavior is different. Here, the analytical value for M_5 is only for the smallest Reynolds number under investigation in fairly good agreement with the numerical calculation. This is not very suprising, because Eq. (4.68) should be correct for large eigenvalues only. (This can be proven by calculating for example M_{40} numerically and by the asymptotic expression. For these values only a deviation between the numerically predicted values and the asymptotic approximation of about 0.5% is found). The analytically predicted values for M_{10} are found to be in fair agreement with the numerical predictions. Instead, for Pr = 0.7, larger deviations can be noticed. However, this is not such a problem, because the length of the thermal entrance region is decreasing with increasing Prandtl numbers and therefore, the impact of the higher eigenvalues and constants on the distribution of the Nusselt number in the thermal entrance region is not so important as for lower Prandtl numbers.

After showing the maximum deviations between the numerical calculations and the asymptotic expressions derived by Sleicher et al. (1970), it is now interesting to see how much the asymptotic analysis can be improved, if the value of β_2 is adapted for $j = 5^5$.

Table 4.3 shows a comparison between numerically predicted values for M_{10}, M_{15} and M_{20} for the two Reynolds numbers $Re_D = 50000$ and $Re_D = 100000$. The values for $Re_D = 10000$ are not shown, because the agreement for these low Reynolds number was already good, even so β_2 was set to zero. As it can be seen from Table 4.3, the agreement between the numerically predicted constants and the ones from the asymptotic approximation is generally quite good. It can also be seen that matching the values for $j = 5$ results in a great improvement in the accuracy. The only values where the agreement is still not fully satisfactory, are the ones for Pr = 0.7 and $Re_D = 100000$. However, for this value of the Prandtl number, the distribution of Nu / Nu_∞ is not much influenced by the Reynolds number (see for example Kays and Crawford (1993)). Therefore, the distribution for Nu / Nu_∞ could be approximated by the one for $Re_D = 50000$, where Nu_∞ has to be taken for the correct Reynolds number.

---

[5] Sleicher et al (1970) used $j = 3$ for adapting their asymptotic results to numerically predicted values for the constants. However, a numerical study showed that matching the asymptotic results for $j = 3$ leads to less accurate results compared to the choice $j = 5$. The reason for this is that one is using the asymptotic formulas in a region ($j = 3$) where they are not valid.

Table 4.1: Comparison between numerically predicted eigenvalues and the asymptotic approximation for $\beta_2 = 0$.

	λ_5	λ_{10}	λ_{20}
Pr = 0.004:		Re$_D$ = 20000	
Numerical	17.4214	32.6512	63.1556
Analytical	17.3215	32.5907	63.1363
		Re$_D$ = 50000	
Numerical	17.8378	33.3973	64.5417
Analytical	17.6861	33.2766	64.4650
		Re$_D$ = 100000	
Numerical	18.5022	34.6224	66.8782
Analytical	18.3134	34.4568	66.7514
Pr = 0.02:		Re$_D$ = 20000	
Numerical	17.8526	33.4928	64.8291
Analytical	17.7860	33.4646	64.8294
		Re$_D$ = 50000	
Numerical	21.5694	40.3435	77.9191
Analytical	21.3428	40.1568	77.7938
		Re$_D$ = 100000	
Numerical	25.7941	48.1753	92.9565
Analytical	25.4343	47.8550	92.7071
Pr = 0.2:		Re$_D$ = 20000	
Numerical	30.3566	56.7416	109.5787
Analytical	29.9993	56.4441	109.3464
		Re$_D$ = 50000	
Numerical	59.0931	110.2750	212.3409
Analytical	57.9515	109.0365	211.2310
		Re$_D$ = 100000	
Numerical	80.9672	151.2447	291.1758
Analytical	79.3562	149.3097	289.2503
Pr = 0.7:		Re$_D$ = 20000	
Numerical	58.2822	108.1118	207.8779
Analytical	56.6920	106.6666	206.6399
		Re$_D$ = 50000	
Numerical	122.3031	227.8361	437.2618
Analytical	118.8246	223.5699	433.1108
		Re$_D$ = 100000	
Numerical	166.9422	312.3776	600.0760
Analytical	162.7946	306.2999	593.3795

Table 4.2: Comparison between numerically predicted constants M_j and the asymptotic approximation for $\beta_2 = 0$.

	$-M_5$	$-M_{10}$	$-M_{20}$
Pr = 0.004:		Re$_D$ = 20000	
Numerical	1.4742	1.2873	1.0589
Analytical	1.5135	1.2265	0.9841
		Re$_D$ = 50000	
Numerical	1.6047	1.4745	1.2972
Analytical	1.9278	1.5623	1.2534
		Re$_D$ = 100000	
Numerical	1.7010	1.5969	1.4591
Analytical	2.3599	1.9124	1.5344
Pr = 0.02:		Re$_D$ = 20000	
Numerical	1.3670	1.1360	0.8951
Analytical	1.3088	1.0607	0.8510
		Re$_D$ = 50000	
Numerical	1.9253	1.7012	1.4829
Analytical	2.3303	1.7708	1.4207
		Re$_D$ = 100000	
Numerical	2.2782	2.0945	1.8812
Analytical	2.9376	2.3807	1.9010
Pr = 0.2:		Re$_D$ = 20000	
Numerical	1.8241	1.5246	1.2175
Analytical	1.8546	1.5029	1.2058
		Re$_D$ = 50000	
Numerical	3.6070	3.2018	2.7545
Analytical	4.2530	3.4466	2.7652
		Re$_D$ = 100000	
Numerical	5.1689	4.3931	3.9156
Analytical	6.2727	5.0834	4.0784
Pr = 0.7:		Re$_D$ = 20000	
Numerical	2.4726	2.2432	1.8339
Analytical	2.8348	2.2973	1.8431
		Re$_D$ = 50000	
Numerical	4.3090	4.5051	4.3173
Analytical	6.8644	5.5629	4.4631
		Re$_D$ = 100000	
Numerical	5.9864	5.6555	5.9483
Analytical	10.1275	8.2073	6.5848

Table 4.3: Comparison between numerically predicted constants M_j and the asymptotic approximation for a value of β_2 so that M_5 is equal .

	$-M_{10}$	$-M_{15}$	$-M_{20}$
Pr = 0.004:		$Re_D = 50000$	
Numerical	1.4745	1.3769	1.2972
Analytical	1.4711	1.3332	1.2301
		$Re_D = 100000$	
Numerical	1.5969	1.5214	1.4869
Analytical	1.7264	1.5982	1.4869
Pr = 0.02:		$Re_D = 50000$	
Numerical	1.7012	1.5804	1.4829
Analytical	1.6832	1.5183	1.3984
		$Re_D = 100000$	
Numerical	2.0945	1.9751	1.8812
Analytical	2.0945	2.0101	1.8625
Pr = 0.2:		$Re_D = 50000$	
Numerical	3.2018	2.9606	2.7545
Analytical	3.2643	2.9497	2.7187
		$Re_D = 100000$	
Numerical	4.3931	4.1307	3.9156
Analytical	4.7718	4.3312	3.9989
Pr = 0.7:		$Re_D = 50000$	
Numerical	4.5051	4.5116	4.3173
Analytical	4.8415	4.5670	4.2790
		$Re_D = 100000$	
Numerical	5.6555	5.9204	5.9483
Analytical	7.0383	6.6904	6.2864

Because of the limited computer power available in the 1970s, the comparison shown here between numerical calculations and the asymptotic expansions has hardly been possible at the time when the asymptotic method has been developed (Sleicher et al (1970)). However, the comparison elucidates very nicely that only the first five eigenvalues need to be calculated numerically. The higher eigenvalues and related constants can than be obtained quite accurately from the asymptotic approximations. The here shown comparisons also give high confidence into the asymptotic approximations developed by Sleicher et al. (1970).

4.2 Heat Transfer in an Axially Rotating Pipe with Constant Wall Temperature

The shown analytical method for calculating larger eigenvalues and constants is now applied to the case of heat transfer in an axially rotating pipe with a hydrodynamically fully developed, turbulent velocity profile. In order to do this, one has to notice that the correct velocity gradient at the wall is $2V$ for the case of the axially rotating pipe. Sleicher et al. (1970) expressed the velocity gradient at the wall as a function of the friction coefficient for a fully developed turbulent pipe flow without rotation. Using this assumption in the present case would lead to wrong answers for the asymptotic eigenvalues and constants. This is due to the fact that, in the case of the axial rotating pipe, the friction coefficient is influenced by system rotation and the laminarization of the flow field. However, if we apply the definition of V and use the results from Chap. 3 and Appendix B for the hydrodynamically fully developed pipe flow in an axially rotating pipe, the correct asymptotic constants and eigenvalues can be predicted. It should also be mentioned here that the constant β_1 attains the value 5/36 as long as the velocity gradient at the wall is assumed to be linear. This is, of course, also a suitable choice for the flow in an axially rotating pipe.

In Chap. 3.3, the heat transfer in an axially rotating pipe has been discussed extensively. However, it is noteworthy mentioning that the laminarization of the flow in the pipe increases with increasing values of the rotation rate $N = w_w / \bar{u} = \mathrm{Re}_\varphi / \mathrm{Re}_D$. This leads to decreasing values of the Nusselt number for increasing values of N.

In a first attempt, the value of β_2 in the asymptotic expressions for the eigenvalues and the constants is set to zero in order to see the maximum deviation between numerically predicted values and the ones from the analytical theory. Again, the eigenvalues are predicted by the simple expression $k_j = j + 2/3$ from Eq. (4.65).

The results obtained are reported in Table 4.4, where comparisons are shown for Pr = 0.71 and different values of Re_D and N. It can be seen that the numerically predicted eigenvalues are in very good agreement with the analytical predictions, even for larger values of the rotation rate. This result is at a first glance somewhat surprisingly, because for larger N, the flow laminarizes and finally for $N \to \infty$ it approaches the behavior of a laminar pipe flow. This means that for $N \to \infty$, $a_2 \to 1$ and $\tilde{u} \to 2\left(1 - \tilde{r}^2\right)$. If we investigate the analytic expression for the larger eigenvalues according to Eq. (4.65), it can be seen that one obtains for λ_j

$$\lambda_j = \frac{1}{\sqrt{2}G}\left(j + \frac{2}{3}\right) \tag{4.69}$$

The quantity G, which appears in Eq. (4.69), is defined by Eq. (4.9). Replacing in the definition of the function G the quantity a_2 by one and the velocity distribution by the one for a hydrodynamically fully developed laminar pipe flow, we obtain

$$G = \frac{1}{\pi} \int_0^1 \sqrt{1 - \tilde{r}^2} \, d\tilde{r} = \frac{1}{2\pi} \arcsin(1) = \frac{1}{4} \qquad (4.70)$$

If we insert this value of G into Eq. (4.69), we obtain from Eq. (4.69) exactly the asymptotic expression of the eigenvalues for laminar pipe flow, Eq. (4.1), which has been predicted by Sellars et al. (1956) (the same behavior can be shown also for the constants M_j).

As it can be seen from Table 4.5, the numerically predicted values for M_j (j =5, 10, 20) for N = 3 and N = 5 are in good agreement with the analytical predictions. However, for N = 1, larger deviations between the numerically predicted values for M_j and the analytical approximation can be noticed. For a Reynolds number of 50000, the largest deviation occurs for N = 1. Here the analytically predicted value for M_5 is about 38% different to the numerically predicted value. For this case, the analytically predicted value of M_{20} is still out by about 15%.

Following the approach of the last paragraph, we can determine again the value of the constant β_2 by enforcing the analytical values of M_5 to be the same as the numerically calculated values. By doing so, it can be noticed from Table 4.5, that the agreement between the numerically predicted constants and the analytical approximations is strongly improved. Exceptions occur for N = 1 and Re_D = 50000, where still larger deviations can be noticed. However, even for this case, the deviations reduce from former 38% for M_{10} to about 24% and for the value of M_{20} from 15% to about 11%.

This shows very nicely that the method of Sleicher et al. (1970) can be used quite easily for other related problems. It should also be kept in mind that, for the case of an axially rotating pipe, usually much more eigenvalues and constants are needed than in the case of a fully turbulent flow without rotation. This is due to the effect of the flow laminarization, which tends to increase dramatically the thermal entrance region (see also Chap. 3). Therefore, the usage of asymptotic values for the eigenvalues and the constants M_j is very beneficial and results in a great simplification for this case.

Table 4.4: Comparison between numerically predicted eigenvalues and the asymptotic approximation for $\beta_2 = 0$ for an axially rotating pipe (Pr = 0.71).

	λ_5	λ_{10}	λ_{20}
$Re_D = 5000$:		N = 1	
Numerical	37.4548	70.1148	135.2362
Analytical	36.9167	69.4903	134.6374
		N = 3	
Numerical	23.9176	44.9527	86.9393
Analytical	23.7882	44.7778	86.7571
		N = 5	
Numerical	18.9059	35.5338	68.7749
Analytical	18.8431	35.4693	68.7218
$Re_D = 10000$:		N = 1	
Numerical	47.6343	89.3193	172.2480
Analytical	46.9711	88.4161	171.3063
		N = 3	
Numerical	28.1981	53.0729	102.6860
Analytical	28.0821	52.8605	102.4171
		N = 5	
Numerical	20.6890	38.9011	75.2881
Analytical	20.6180	38.8104	75.1951
$Re_D = 20000$:		N = 1	
Numerical	61.4338	115.5300	222.8338
Analytical	60.7089	114.2756	221.4090
		N = 3	
Numerical	34.3362	64.7308	125.3610
Analytical	34.2711	64.5103	124.9887
		N = 5	
Numerical	23.3372	43.9092	85.0086
Analytical	23.2725	43.8070	84.8761
$Re_D = 50000$:		N = 1	
Numerical	87.4261	165.2109	319.0354
Analytical	86.8290	163.4429	316.6706
		N = 3	
Numerical	46.4773	87.7990	170.3082
Analytical	46.5653	87.6524	169.8265
		N = 5	
Numerical	29.0058	54.6220	105.8197
Analytical	28.9729	54.5373	105.6660

Table 4.5: Comparison between numerically predicted constants M_j and the asymptotic approximation for $\beta_2 = 0$ for an axially rotating pipe (Pr = 0.71).

	$-M_5$	$-M_{10}$	$-M_{20}$
Re$_D$ = 5000:		N = 1	
Numerical	1.5102	1.3339	1.1157
Analytical	1.7269	1.3986	1.1219
		N = 3	
Numerical	1.0368	0.8485	0.7024
Analytical	1.0918	0.8843	0.7093
		N = 5	
Numerical	0.8287	0.6778	0.5488
Analytical	0.8451	0.8451	0.6844
Re$_D$ = 10000:		N = 1	
Numerical	1.8785	1.6993	1.4754
Analytical	2.3237	1.8820	1.5097
		N = 3	
Numerical	1.2784	1.0065	0.8370
Analytical	1.3429	1.0877	0.8725
		N = 5	
Numerical	0.9212	0.7422	0.6048
Analytical	0.9416	0.7626	0.6118
Re$_D$ = 20000:		N = 1	
Numerical	2.4120	2.1686	1.9620
Analytical	3.2109	2.6006	2.0861
		N = 3	
Numerical	1.7083	1.2660	1.0244
Analytical	1.7340	1.4044	1.1266
		N = 5	
Numerical	1.0992	0.8583	0.6877
Analytical	1.0972	0.8887	0.7129
Re$_D$ = 50000:		N = 1	
Numerical	3.6757	2.9977	2.8559
Analytical	5.0721	4.1080	3.2953
		N = 3	
Numerical	2.8233	1.8908	1.4061
Analytical	2.5829	2.0920	1.6781
		N = 5	
Numerical	1.5958	1.1845	0.8982
Analytical	1.4593	1.1819	0.9481

Table 4.6: Comparison between numerically predicted constants M_j and the asymptotic approximation for a value of β_2 so that M_5 is equal (axial rotating pipe, Pr = 0.71).

	$-M_{10}$	$-M_{15}$	$-M_{20}$
Re$_D$ = 5000:		N = 1	
Numerical	1.3339	1.2104	1.1157
Analytical	1.3376	1.2027	1.1063
		N = 3	
Numerical	0.8485	0.7613	0.7024
Analytical	0.8688	0.7709	0.7054
Re$_D$ = 10000:		N = 1	
Numerical	1.6993	1.5777	1.4754
Analytical	1.7565	1.5987	1.4776
		N = 3	
Numerical	1.0965	0.8999	0.8370
Analytical	1.0695	0.9486	0.8679
Re$_D$ = 20000:		N = 1	
Numerical	2.1686	2.0661	1.9620
Analytical	2.3754	2.1856	2.0286
		N = 3	
Numerical	1.2660	1.1057	1.0243
Analytical	1.3972	1.2323	1.1247
Re$_D$ = 50000:		N = 1	
Numerical	2.9977	2.9370	2.8559
Analytical	3.7143	3.4353	3.1948
		N = 3	
Numerical	1.8908	1.5604	1.4061
Analytical	2.1599	1.8712	1.6954

4.3 Asymptotic Expressions for other Thermal Boundary Conditions

Several authors have used the method of Sleicher et al. (1970) in order to obtain asymptotic expressions for related problems involving turbulent internal flow and other boundary conditions as the constant wall temperature. Since the mathematical approach is nearly identical to the one presented in detail in this chapter, only the resulting expressions for the eigenvalues and the constants are reported here. For the case of a constant heat flux at the pipe wall, Notter and Sleicher (1971) obtained the following expressions for the eigenvalues and the constants

$$k_j = j + \frac{1}{3} - \frac{3^{7/6} G^{2/3} \Gamma(4/3)}{4\pi V^{1/3} \Gamma(2/3)(j+1/3)^{2/3}} \tag{4.71}$$

$$A_j = \frac{(-1)^{j+1} 3^{4/3} G^{1/3} \Gamma(4/3)}{\sqrt{2A} \, \pi V^{1/6} k_j^{4/3}} / \left\{ 1 - \frac{3^{2/3} G^{2/3} \Gamma(4/3)}{4V^{1/3} \Gamma(2/3) k_j^{2/3}} \right\} \tag{4.72}$$

$$A_j \Phi_j(1) = \frac{3^{7/6} 2 \Gamma(4/3)}{\pi V^{1/3} G^{-2/3} \Gamma(2/3)(k_j)^{5/3}} / \left\{ 1 - \frac{3^{2/3} \Gamma(1/3)}{4 G^{-2/3} V^{1/3} \Gamma(2/3)(k_j)^{2/3}} \right\} \tag{4.73}$$

$$\left\{ 1 + \frac{3^{2/3} \Gamma(4/3)}{4V^{1/3} G^{-2/3} \Gamma(2/3)(k_j)^{2/3}} + \frac{1}{k_j} \frac{1}{2\sqrt{3}} \left(\beta_2 \ln(k_j \pi) + \frac{7}{18\pi} \right) \right\}$$

Notter and Sleicher (1971b) compared their analytical results for the eigenvalues and the constants according to Eq. (4.71) and Eq. (4.73) with numerical calculations for $j = 4$ and found good agreement for $Pr \le 0.06$.

The case of a convective cooled wall is a more general case than the two above considered cases of constant wall temperature and constant wall heat flux. For this case, the eigenvalue problem is given by

$$\frac{d}{d\tilde{r}} \left[\tilde{r} a_2(\tilde{r}) \frac{d\Phi_j}{d\tilde{r}} \right] + \tilde{r} \tilde{u}(\tilde{r}) \lambda_j^2 \Phi_j(\tilde{r}) = 0 \tag{4.7}$$

with the boundary conditions

$$\tilde{r} = 0 : \Phi'(0) = 0 \tag{4.74}$$

$$\tilde{r} = 1 : \Phi'(1) + Bi \, \Phi(1) = 0$$

and with the Biot number defined by $Bi = hR/k$. For $Bi \to 0$ one obtains from Eq. (4.74) the constant heat flux case, while for $Bi \to \infty$ one obtains the case of a constant pipe wall. For the case of a convective cooled pipe wall, Sadeghipour et al. (1984) derived asymptotic expressions for the eigenvalues and constants. For their analysis, they assumed a turbulent velocity profile according to

$$\tilde{u} = 2c(1-\tilde{r})^{1/m}, \quad c = \frac{1}{4} \left(\frac{3m+1}{m^2} + 2 \right) \tag{4.75}$$

This means that the following expressions cannot be used for the heat transfer in an axially rotating pipe. However, Eqs. (4.71-4.73) (for the case of a constant wall heat flux) can be used also for the case of an axially rotating pipe, because in these equations the velocity gradient at the wall, $2V$, has been used. This value needs only to be adopted, as it was discussed in the previous section.

The asymptotic expressions by Sadeghipour et al. (1984) are

$$k_j = \frac{(-1)^j}{2k_j\pi}\left(\frac{\omega\cos\varepsilon_1 + \omega\cos\varepsilon_2}{\omega+1}\right)\left(\beta_2\ln\left(k_j\pi\right)+\frac{1-2\gamma^2}{2\pi}\right)+$$

$$j+\frac{1}{2}\left(1+\gamma\frac{\omega-1}{\omega+1}\right) \tag{4.76}$$

where the following abbreviations are used

$$\gamma = \frac{m}{2m+1}, \quad \psi_1 = \frac{\pi}{2}(1+\gamma), \quad \psi_2 = \frac{\pi}{2}(1-\gamma), \tag{4.77}$$
$$\varepsilon_1 = k_j\pi - \psi_1, \quad \varepsilon_2 = k_j\pi - \psi_2$$

$$\omega = \frac{1+\gamma}{1-\gamma}\frac{(\text{Bi}-1/2)}{\left(\sqrt{c}\gamma k_j / G\right)^{2\gamma}} \tag{4.78}$$

$$A_j\Phi_j(1) = \frac{4\,\text{Bi}}{\pi k_j}\left(2\sqrt{\left|\frac{2m+1}{m+1}\frac{G}{k_j}\right|}\right)^{2\gamma}\frac{\Gamma(1+\gamma)}{\Gamma(1-\gamma)}$$

$$\frac{-\sin\varepsilon_1 + \dfrac{1}{2k_j}\left(\dfrac{1-2\gamma^2}{2\pi}+\beta_2\ln\left(k_j\pi\right)\right)\cos\varepsilon_1}{\cos\varepsilon_2 + \omega\cos\varepsilon_1 + \dfrac{a_{11}}{2k_j}+\dfrac{a_{12}}{2\pi k_j^2}} \tag{4.79}$$

$$a_{11} = \left(-\frac{1+2v^2}{2\pi}+\beta_2\ln\left(k_j\pi\right)\right)(\sin\varepsilon_2 + \omega\sin\varepsilon_1)+\frac{4v}{\pi}\sin\varepsilon_2 \tag{4.80}$$

$$a_{12} = \left[-\beta_2+\frac{3}{2}\left(\frac{1-2v^2}{2\pi}+\beta_2\ln\left(k_j\pi\right)\right)\right](\cos\varepsilon_2 + \omega\cos\varepsilon_1)$$
$$-2v\left(\frac{1-2v^2}{\pi}+\beta_2\ln\left(k_j\pi\right)\right)\cos\varepsilon_2 \tag{4.81}$$

Problems

4-1. Consider the heat transfer for laminar pipe flow. The eigenvalue problem for the heat transfer in the thermal entrance region is given by

$$\frac{d}{d\tilde{r}}\left[\tilde{r}\frac{d\Phi_j}{d\tilde{r}}\right] + \tilde{r}2\left(1-\tilde{r}^2\right)\Phi_j = 0$$

with the boundary conditions

$$\tilde{r} = 0 : \frac{d\Phi}{d\tilde{r}} = 0$$

$$\tilde{r} = 1 : \Phi = 0$$

Show by introducing the quantities defined by Eq. (4.9-4.10) that the problem can be reduced to the form given by Eqs. (4.11-4.12).

4.2. Consider the heat transfer in a parallel plate channel with distance $2h$ between the two plates. The flow is laminar and hydrodynamically fully developed. The velocity profile is given by

$$\tilde{u} = \frac{3}{2}\left(1-\left(\frac{y}{h}\right)^2\right)$$

The problem can be described by the following partial differential equation

$$\tilde{u}\frac{\partial\Theta}{\partial\tilde{x}} = \frac{\partial^2\Theta}{\partial y^2}$$

and the boundary conditions

$$\tilde{x} = 0 : \Theta = 1$$

$$\tilde{y} = 0 : \frac{\partial\Theta}{\partial y} = 0$$

$$\tilde{y} = 1 : \Theta = 0$$

Use the FORTRAN program reported in Appendix C to calculate the eigenvalues for this problem. Plot the first 100 eigenvalues λ_j^2 as a function of j. Provide an equation for large eigenvalues.

5 Heat Transfer in Duct Flows for Small Peclet Numbers (Elliptic Problems)

In Chapters 3-4, we have been concerned with heat transfer in hydrodynamically fully developed flow in a pipe or a planar channel. There we made the assumption that the axial heat conduction within the fluid can be ignored. This assumption implies that the Peclet number $\left(\mathrm{Pe}_D = \mathrm{Re}_D \mathrm{Pr}\right)$ for the problem under consideration has to be large enough. If one neglects the axial heat conduction term in the energy equation, the partial differential equation (3.5) changes its type from elliptic to parabolic. This means that the original energy equation (3.5) is elliptic, whereas the simplified equation (e.g. Eq. (3.14)) is parabolic. This has the consequence that a boundary condition for the parabolic problem has to be prescribed at the entrance into the heated section at $x = 0$ (see Fig. 3.1). This shows clearly, that for parabolic problems no information can be transferred upstream of the heated zone into the unheated zone $\left(x < 0\right)$. As stated above, this is a good approximation, if the Peclet number is large $\left(\mathrm{Pe}_D > 100\right)$ and thus axial heat conduction effects in the fluid can be neglected. However, for several technical applications the Peclet number is small, so that the axial heat conduction within the fluid has to be taken into account. A typical situation for this case is a compact heat exchanger using liquid metals as working fluid. For liquid metals, the Prandtl number is very low ($0.001 < \mathrm{Pr} < 0.1$) and therefore the Peclet number may attain very small values. In addition to the above mentioned example, there exist other types of applications, where one has to include the effect of axial heat conduction within the fluid, even for larger Peclet numbers. These are applications where the heating section is restricted in length as, for example, in cooling systems for electronic components or in chemical process lines.

The present chapter concentrates on the analytical treatment of such problems. Because of their similarity to the Graetz problems considered in the previous chapters, the problems under consideration are often denoted as "extended Graetz problems". However, the mathematical treatment of these problems is much more complicated than the one for the parabolic problems as discussed later in more detail. In the following section, a short literature overview is given in order to show the work which has been done in the past on this subject. Many investigations have been carried out which deal with the solution of the extended Graetz problem for thermally developing laminar flow in a pipe or in a parallel plate channel. Good literature reviews on this subject are given by Shah and London (1978) and by Reed (1987). Many of the solutions for the extended Graetz problem, provided in the above review articles, are based on the fundamental assumption that the solution of the problem has the same functional form as in the Graetz problem without axial heat conduction. This approach results in a non self-adjoint eigenvalue problem with eigenvalues that could, at least in principle, be complex and eigenvectors that could be incomplete. Several strategies have been developed in the past to overcome this problem. Hsu (1971b) constructed the solution of the problem from two independent series solutions for $x < 0$ and $x > 0$. Both, the temperature distribution and the temperature gradient were then matched at $x = 0$ by constructing a pair of orthonormal functions from the non orthogonal eigenfunctions by using the Gram - Schmidt - orthonormalization procedure (see Broman (1970) for an explanation of the method). Hence this method is clearly plagued with the uncertainties arising from an expansion in terms of eigenfunctions belonging to a non self-adjoint operator (in order to obtain a solution by the Gram-Schmidt orthonormalization procedure a coupled system of equations has to be solved). Despite these disadvantages, the technique of deriving the solution from two independent solutions for the different regions ($x < 0$, $x > 0$) and matching the resulting expressions for $x = 0$ has been used by several authors in the past. The reader is referred to the work of Hsu (1970), Bayazitoglu and Özisik (1980), Vick at al. (1980) and Vick and Özisik (1981). Jones (1971) used a Laplace transform technique for the elliptic energy equation for laminar pipe flow. A completely different approach was presented by Papoutsakis et al. (1980a, 1980b). They showed that it is possible to derive an entirely analytical solution to the extended Graetz problem for constant wall temperature and constant wall heat flux boundary conditions for laminar pipe flow. Their solution is based on a self-adjoint formalism resulting from a decomposition of the convective diffusion equation into a pair of first order partial differential equations. The method is based on the work of Ramkrishna and Amundson (1979a, 1979b). Later, an alternative approach for obtaining an analytical solution of the extended Graetz problem was presented by Ebadian and Zhang (1989). They used a Fourier transform of the temperature field and expanded the coefficients of the transformed temperature in terms of the Peclet number. This approach resulted in a set of ordinary differential equations, which have to be solved successively.

In addition, several investigations have been carried out in the past concerning the extended Graetz problem in a parallel plate channel. Deavours (1974) presented an analytical solution for the extended Graetz problem by decomposing the eigenvalue problem for the parallel plate channel into a system of ordinary differential equations for which he proved the orthogonality of the eigenfunctions. Weigand et al. (1993) investigated liquid solidification in a parallel plate channel subjected to laminar internal flow. They applied a regular perturbation expansion to the energy equation. The zero order problem was formally identical to the extended Graetz problem in a parallel plate channel. For the solution of the zero order problem they followed the method of Papoutsakis et al. (1980a) and obtained an analytical solution.

Because of the difficulties involved in solving analytically the extended Graetz problems, researchers focused very early on the usage of numerical solution methods. Hennecke (1968) calculated numerically, by using a finite difference method, the distribution of the temperature and the Nusselt number for laminar pipe flow. Results are obtained for the two different wall boundary conditions: a constant wall temperature and a constant wall heat flux. Hennecke showed that the axial heat conduction effect is more pronounced for the constant wall temperature situation. In addition, he was able to show that the influence of axial heat conduction on the temperature field and the distribution of the Nusselt number can be ignored for $Pe_D > 100$. These results are in good agreement with the investigation of Acrivos (1980), who showed analytically that the region of influence for axial heat conduction, scaled by the pipe radius, is of the order of Pe_D^{-1}. Nguyen (1992) and Bilir (1992) investigated numerically the heat transfer for thermally developing flow in a circular pipe and in a planar channel. Nguyen (1992) derived from his computational results accurate engineering correlations for the Peclet number effect on the local Nusselt number in the thermal entry region.

Although axial heat conduction can be ignored for turbulent convection in ordinary fluids and gases, with liquid metals this may not always be justified. In fact, due to the very low Prandtl numbers for liquid metals, the Peclet number can be as small as three in turbulent duct flows. Reviews on the convective heat transfer in liquid metals can be found in Reed (1987), Stein (1966) and Kays and Crawford (1993). Lee (1982) studied the extended Graetz problem in turbulent pipe flow. He found that, for Peclet numbers smaller than 100, axial heat conduction in the fluid becomes important in the thermal entrance region. He investigated a pipe which was insulated for $x \leq 0$ and had a uniform wall temperature for $x > 0$. Lee (1982) used the method of Hsu (1971b) to obtain a series solution for the problem. In this work, the variation of the Nusselt number for fully-developed flow was shown for several Peclet and Prandtl numbers. In addition, the error in Nusselt number resulting from neglecting the axial heat conduction effect was presented. This error was found to be as high as 40% for a Peclet number of 5.

For the case of turbulent flow inside a parallel plate channel, the effect of axial heat conduction within the fluid was studied by Faggiani and Gori (1980). They solved numerically the energy equation for a constant heat flux boundary condition. Weigand (1996) extended the method of Papoutsakis et al. (1980a) to solve the extended Graetz problem for turbulent flow in a pipe and in a planar channel. Like the solution given by Papoutsakis et al. (1980a) for laminar internal flow, the solution by Weigand (1996) has not been plagued by the uncertainties arising from expansion in terms of eigenfunctions, which belong to a non self-adjoint eigenvalue system. Later, Weigand et al. (1997) and Weigand and Wrona (2003) extended the method also for heat transfer in concentric annuli.

Nearly all studies mentioned here considered only a heating section semi-infinite in size. Hennecke (1968) was one of the few who calculated numerically the temperature distribution at the end of the heating section for laminar internal flow. However, he investigated this problem only for a very large heating section. The problem was solved later analytically by Papoutsakis et al. (1980b) for laminar pipe flow and a heating zone of finite length. They derived expressions for the Nusselt number and the bulk-temperature, but did not show, however, any distributions for these quantities. Only temperature profiles in the fluid were shown.

Weigand et al. (2001) calculated the effect of a changing length of the heating zone for laminar and turbulent internal flow and for different Peclet numbers. One of the main results of their study was that, for finite heated sections, the axial heat conduction effect depends also on the length of the heated zone. Axial heat conduction may, therefore, be important also for Peclet numbers larger than 100 if this zone is short.

The present chapter shows an analytical solution method for the extended Graetz problem. The method is based on the approach by Papoutsakis (1980a, 1980b) and can be used for a lot of related problems. First, the method is explained for a duct with a semi-infinite heating section (with constant wall temperature or constant wall heat flux) to highlight the basic ideas behind the solution approach. Later applications of the method to piecewise heated ducts are explained. The goal of this chapter is to show that the derived final solutions for the extended Graetz problem are as simple as the ones for the parabolic problems discussed in Chap. 3.

5.1 Heat Transfer for Constant Wall Temperatures for x≤ 0 and x > 0

We start our considerations for a hydrodynamically fully developed flow with a low Peclet number and constant wall temperature. As for the analysis in Chap. 3, it is assumed that the velocity profile is fully developed. This means that the vertical velocity component v is zero and the axial velocity component is only a function of the coordinate orthogonal to the main flow direction. The geometry under consideration and the used coordinate system is shown in Fig. 5.1.

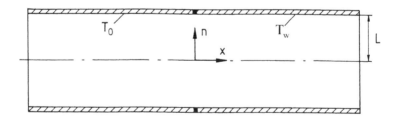

Fig. 5.1: Geometry and coordinate system (Weigand (1996))

If we assume constant fluid properties, the energy equation is given by Eq. (3.5) and Eq. (3.6)

Pipe

$$\rho c_p u(r) \frac{\partial T}{\partial x} = \frac{1}{r} \frac{\partial}{\partial r} \left[r \left(k \frac{\partial T}{\partial r} - \rho c_p \overline{v'T'} \right) \right] + \frac{\partial}{\partial x} \left[k \frac{\partial T}{\partial x} - \rho c_p \overline{u'T'} \right] \qquad (3.5)$$

Planar Channel

$$\rho c_p u(y) \frac{\partial T}{\partial x} = \frac{\partial}{\partial y} \left[k \frac{\partial T}{\partial y} - \rho c_p \overline{v'T'} \right] + \frac{\partial}{\partial x} \left[k \frac{\partial T}{\partial x} - \rho c_p \overline{u'T'} \right] \qquad (3.6)$$

If one analyses Eq. (3.5) or (3.6) for laminar flow ($\overline{u'T'} = 0$, $\overline{v'T'} = 0$), one finds immediately that the equations are elliptic in nature. This is due to the axial heat conduction effect within the flow, which is represented by the second term on the right hand side of the equations. For turbulent flow, the nature of the equation will also depend on the turbulent heat fluxes. The turbulent heat fluxes $-\rho c_p \overline{v'T'}$ and $-\rho c_p \overline{u'T'}$ have to be modeled. This can be done for example by using a simple eddy viscosity model[1]:

$$-\overline{v'T'} = \varepsilon_{hy} \frac{\partial T}{\partial y} \text{ (planar channel)}, \qquad -\overline{v'T'} = \varepsilon_{hr} \frac{\partial T}{\partial r} \text{ (pipe)} \qquad (3.7)$$

$$-\overline{u'T'} = \varepsilon_{hx} \frac{\partial T}{\partial x} \text{ (planar channel and pipe)} \qquad (3.8)$$

where ε_{hy}, ε_{hr} and ε_{hx} are only functions of the coordinate orthogonal to the flow direction. Inserting Eqs. (3.7-3.8) into Eq. (3.5) and Eq. (3.6) results in

Pipe

$$\rho c_p u(r) \frac{\partial T}{\partial x} = \frac{1}{r} \frac{\partial}{\partial r} \left[r \left(k + \rho c_p \varepsilon_{hr} \right) \frac{\partial T}{\partial r} \right] + \frac{\partial}{\partial x} \left[\left(k + \rho c_p \varepsilon_{hx} \right) \frac{\partial T}{\partial x} \right] \qquad (3.9)$$

---

[1] More advanced models can also be used, which assume the stream-wise component of the turbulent heat flux to be also a function of the cross-stream temperature gradient and vice versa (see Batchelor (1949), So and Sommer (1996) and Sommer (1994)). Weigand et al. (2002) conducted a numerical analysis of the energy equation (3.6) and found the above mentioned simple model to be in perfect agreement with more complicated models for the range of parameters considered here.

Planar Channel

$$\rho c_p u(y) \frac{\partial T}{\partial x} = \frac{\partial}{\partial y}\left[\left(k + \rho c_p \varepsilon_{hy}\right)\frac{\partial T}{\partial y}\right] + \frac{\partial}{\partial x}\left[\left(k + \rho c_p \varepsilon_{hx}\right)\frac{\partial T}{\partial x}\right] \qquad (3.10)$$

If one replaces in Eqs. (3.9-3.10) the heat diffusivities ε_{hy}, ε_{hr} and ε_{hx} by a turbulent Prandtl number according to Eq. (3.18), one obtains

Pipe

$$\rho c_p u(r) \frac{\partial T}{\partial x} = \frac{1}{r}\frac{\partial}{\partial r}\left[r\left(k + \rho c_p \frac{\varepsilon_m}{Pr_t}\right)\frac{\partial T}{\partial r}\right] + \frac{\partial}{\partial x}\left[\left(k + \rho c_p \frac{\varepsilon_m}{Pr_t}\frac{\varepsilon_{hx}}{\varepsilon_{hr}}\right)\frac{\partial T}{\partial x}\right] \qquad (5.1)$$

Planar Channel

$$\rho c_p u(y) \frac{\partial T}{\partial x} = \frac{\partial}{\partial y}\left[\left(k + \rho c_p \frac{\varepsilon_m}{Pr_t}\right)\frac{\partial T}{\partial y}\right] + \frac{\partial}{\partial x}\left[\left(k + \rho c_p \frac{\varepsilon_m}{Pr_t}\frac{\varepsilon_{hx}}{\varepsilon_{hy}}\right)\frac{\partial T}{\partial x}\right] \qquad (5.2)$$

Eqs. (5.1-5.2) can be combined to one equation if we introduce, like in Chap. 3, a vertical coordinate n which is equal to y for a planar channel and equal to r for pipe flows. This results in

$$\rho c_p u(n) \frac{\partial T}{\partial x} = \frac{1}{r^F}\frac{\partial}{\partial n}\left[r^F\left(k + \rho c_p \frac{\varepsilon_m}{Pr_t}\right)\frac{\partial T}{\partial n}\right] + \frac{\partial}{\partial x}\left[\left(k + \rho c_p \frac{\varepsilon_m}{Pr_t}\right)\frac{\partial T}{\partial x}\right] \qquad (5.3)$$

where the superscript F specifies the geometry and has to be set to zero for a planar channel and equal to one for pipe flow. The boundary conditions for Eq. (5.3) are

$$n = L : T = T_0 , \ x \le 0 \qquad (5.4)$$
$$T = T_W , x > 0$$
$$n = 0 : \partial T / \partial n = 0$$
$$\lim_{x \to -\infty} T = T_0 , \qquad \lim_{x \to +\infty} T = T_W$$

Introducing the following dimensionless quantities into the above equations

$$\Theta = \frac{T - T_W}{T_0 - T_W} , \ \tilde{r} = \frac{r}{R} , \ \tilde{x} = \frac{x}{L}\frac{1}{Re_L Pr} , \ \tilde{n} = \frac{n}{L} , \ Pr = \frac{\mu c_p}{k} , \ Re_L = \frac{\bar{u} L}{v} , \qquad (5.5)$$

$$\tilde{\varepsilon}_m = \frac{\varepsilon_m}{v} , \ \tilde{u} = \frac{u}{\bar{u}} , \ Pe_L = Re_L Pr$$

where the length scale $L = R$ for a pipe flow and $L = h$ for the flow in a planar channel, one obtains

$$\tilde{u}(\tilde{n})\frac{\partial \Theta}{\partial \tilde{x}} = \frac{1}{Pe_L^2}\frac{\partial}{\partial \tilde{x}}\left[a_1(\tilde{n})\frac{\partial \Theta}{\partial \tilde{x}}\right] + \frac{1}{\tilde{r}^F}\frac{\partial}{\partial \tilde{n}}\left[\tilde{r}^F a_2(\tilde{n})\frac{\partial \Theta}{\partial \tilde{n}}\right] \tag{5.6}$$

with the boundary conditions

$$\tilde{n}=1: \quad \Theta=1, \; \tilde{x}\leq 0 \tag{5.7}$$
$$\Theta=0, \; \tilde{x}>0$$
$$\tilde{n}=0: \partial \Theta/\partial \tilde{n}=0$$
$$\lim_{\tilde{x}\to-\infty}\Theta=0\,, \qquad \lim_{\tilde{x}\to+\infty}\Theta=1$$

The functions $a_1(\tilde{n})$ and $a_2(\tilde{n})$ are given by

$$a_1(\tilde{n})=1+\frac{Pr}{Pr_t}\tilde{\varepsilon}_m\,, \qquad a_2(\tilde{n})=1+\frac{Pr}{Pr_t}\left(\frac{\varepsilon_{hx}}{\varepsilon_{hn}}\right)\tilde{\varepsilon}_m \tag{5.8}$$

Papoutsakis et al. (1980a) showed that it is possible to solve Eq. (5.6) for laminar pipe flow ($a_1 = a_2 = 1$, $F = 1$) by decomposing the elliptic partial differential equation into a pair of first order partial differential equations. The ensuing procedure for solving the extended turbulent Graetz problem given by Eqs. (5.6-5.7) follows the method of Papoutsakis et al. (1980a) for deriving the solution of the more general problem where a_1 and a_2 are functions of $\tilde{n}$.

In order to obtain a solution for the turbulent Graetz problem given by Eqs. (5.6-5.7), one can proceed in the following way: First, we define a function $E(\tilde{x},\tilde{n})$ which may be called the axial energy flow through a cross - sectional area of height $\tilde{n}$. This function is given by

$$E = \int_0^{\tilde{n}}\left[\tilde{u}\Theta-\frac{1}{Pe_L^2}a_1(\overline{n})\frac{\partial \Theta}{\partial \tilde{x}}\right]\tilde{r}^F d\overline{n} \tag{5.9}$$

If one introduces the axial energy flow according to Eq. (5.9) into the energy equation (5.6), one obtains the following system of first order partial differential equations

$$\frac{\partial}{\partial \tilde{x}}\vec{S}(\tilde{x},\tilde{n})=\underset{\sim}{L}\,\vec{S}(\tilde{x},\tilde{n}) \tag{5.10}$$

where the solution vector $\vec{S}$ and the matrix operator $\underset{\sim}{L}$ are defined by

$$\vec{S} = \begin{bmatrix} \Theta(\tilde{x}, \tilde{n}) \\ E(\tilde{x}, \tilde{n}) \end{bmatrix}, \qquad L = \begin{bmatrix} \dfrac{Pe_L\, \tilde{u}(\tilde{n})}{a_1(\tilde{n})} & -\dfrac{Pe_L^2}{\tilde{r}^F\, a_1(\tilde{n})} \dfrac{\partial}{\partial \tilde{n}} \\[4mm] \tilde{r}^F\, a_2(\tilde{n}) \dfrac{\partial}{\partial \tilde{n}} & 0 \end{bmatrix} \qquad (5.11)$$

The boundary conditions for the function E can be obtained directly from its definition, Eq. (5.9), by using Eq. (5.7). This results in

$$\lim_{\tilde{x} \to -\infty} E = \int_0^{\tilde{n}} \tilde{u}\, \tilde{r}^F\, d\tilde{n}, \qquad \lim_{\tilde{x} \to +\infty} E = 0 \qquad \tilde{n} = 0 : E = 0 \qquad (5.12)$$

Before solving Eq. (5.10), some interesting features of the operator L and the corresponding eigenvalue problem for this equation should be presented. The most remarkable aspect of L is that it gives rise to a self-adjoint problem even though the original convective diffusion operator is non self-adjoint. This fact is of course dependent on the sort of inner product between two vectors, which will be used. This means that we have to find an inner product which is symmetric. Papoutsakis et al. (1980a) obtained such an inner product for a laminar pipe flow. It is now possible to extend this work for laminar and turbulent duct flows. Let us define the following inner product for two vectors

$$\vec{\Phi} = \begin{bmatrix} \Phi_1(\tilde{n}) \\ \Phi_2(\tilde{n}) \end{bmatrix} \qquad , \vec{\Upsilon} = \begin{bmatrix} \Upsilon_1(\tilde{n}) \\ \Upsilon_2(\tilde{n}) \end{bmatrix} \qquad (5.13)$$

by

$$\langle \vec{\Phi}, \vec{\Upsilon} \rangle = \int_0^1 \left[\frac{a_1(\tilde{n})\, \tilde{r}^F}{Pe_L^2} \Phi_1(\tilde{n})\, \Upsilon_1(\tilde{n}) + \frac{1}{a_2(\tilde{n})\, \tilde{r}^F} \Phi_2(\tilde{n})\, \Upsilon_2(\tilde{n}) \right] d\tilde{n} \qquad (5.14)$$

The inner product of the two vectors $\vec{\Phi}, \vec{\Upsilon}$ incorporates the two functions a_1, a_2 which contain the heat diffusivity for the turbulent flow. In case of a laminar pipe flow, the two functions will be equal to one. L belongs to the following domain

$$D(L) = \left\{ \vec{\Phi} \in H : L\, \vec{\Phi}\ (\text{exists and}) \in H, \Phi_1(1) = \Phi_2(0) = 0 \right\} \qquad (5.15)$$

Then it can be shown that the matrix operator L is a symmetric operator in the Hilbert space H. This means that

$$\langle \vec{\Phi}, L\vec{\Upsilon} \rangle = \langle L\vec{\Phi}, \vec{\Upsilon} \rangle \qquad (5.16)$$

Eq. (5.16) can be proven by inserting the expressions $L\vec{\Phi}$ and $L\vec{\Upsilon}$ into the definition of the inner product according to Eq. (5.14). This results in

$$\left\langle \vec{\Phi}, \underset{\sim}{L}\vec{\Upsilon} \right\rangle = \int\limits_0^1 \left[\frac{\tilde{u}\,\tilde{r}^F}{Pe_L}\Phi_1(\tilde{n})\,\Upsilon_1(\tilde{n}) - \Phi_1(\tilde{n})\Upsilon_2'(\tilde{n}) + \Phi_2(\tilde{n})\,\Upsilon_1'(\tilde{n}) \right] d\tilde{n} \qquad (5.17)$$

$$\left\langle \underset{\sim}{L}\vec{\Phi}, \vec{\Upsilon} \right\rangle = \int\limits_0^1 \left[\frac{\tilde{u}\,\tilde{r}^F}{Pe_L}\Phi_1(\tilde{n})\,\Upsilon_1(\tilde{n}) - \Phi_2'(\tilde{n})\Upsilon_1(\tilde{n}) + \Phi_1'(\tilde{n})\,\Upsilon_2(\tilde{n}) \right] d\tilde{n} \qquad (5.18)$$

Substracting Eq. (5.18) from Eq. (5.17) results after integration in

$$\left\langle \vec{\Phi}, \underset{\sim}{L}\vec{\Upsilon} \right\rangle - \left\langle \underset{\sim}{L}\vec{\Phi}, \vec{\Upsilon} \right\rangle = \Phi_2(1)\,\Upsilon_1(1) - \Phi_2(0)\,\Upsilon_1(0) \qquad (5.19)$$
$$- \Phi_1(1)\,\Upsilon_2(1) + \Phi_1(0)\,\Upsilon_2(0)$$

The resulting expression on the right hand side of Eq. (5.19) is zero, if the vectors $\vec{\Phi}$ and $\vec{\Upsilon}$ satisfy the conditions given in Eq. (5.15).

The associated eigenvalue problem to Eq. (5.10) is given by

$$\underset{\sim}{L}\vec{\Phi}_j = \lambda_j \vec{\Phi}_j \qquad (5.20)$$

In this equation $\vec{\Phi}_j$ denotes the eigenfunction corresponding to the eigenvalue λ_j. If we introduce the definition of the matrix operator $\underset{\sim}{L}$ into this equation, the following eigenvalue problem is obtained

$$Pe_L^2 \left[\frac{\tilde{u}}{a_1(\tilde{n})}\Phi_{j1} - \frac{1}{\tilde{r}^F a_1(\tilde{n})}\Phi_{j2}' \right] = \lambda_j\,\Phi_{j1} \qquad (5.21)$$

$$\tilde{r}^F a_2(\tilde{n})\Phi_{j1}' = \lambda_j\,\Phi_{j2} \qquad (5.22)$$

From the above system of ordinary differential equations, one differential equation for the first component of the eigenvector Φ_{j1} can be easily obtained by eliminating Φ_{j2}. This results in

$$\left[\tilde{r}^F a_2(\tilde{n})\Phi_{j1}' \right]' + \tilde{r}^F \left[\frac{\lambda_j\,a_1(\tilde{n})}{Pe_L^2} - \tilde{u} \right]\lambda_j\,\Phi_{j1} = 0 \qquad (5.23)$$

Eq. (5.23) has to be solved together with the homogenous boundary conditions

$$\Phi_{j1}'(0)=0 \qquad\qquad , \Phi_{j1}(1)=0 \qquad (5.24)$$

Usually Eq. (5.23) is solved numerically with the procedure described in Appendix C. The method used is a "shooting method", where the differential equation (5.23) is integrated from the centreline of the duct ($\tilde{n} = 0$) to the wall and then the boundary condition at the wall $\Phi_{j1}(1) = 0$ is checked for an assumed value of λ_j.

Therefore, an additional condition for $\tilde{n} = 0$ is required. A possible condition is that

$$\Phi_{j1}(0) = 1 \tag{5.25}$$

This condition is a normalizing condition for the eigenfunctions Φ_{j1}. Eq. (5.23) shows clearly that the eigenvalues appear as λ_j and λ_j^2 in the equation. This is different to the normal Graetz problem (see Eq. (3.25)). For the eigenvalue problem according to Eq. (5.20) it is possible to show that it is self-adjoint and semi-definite (see Sauer and Szabo (1969)). Therefore, all eigenvalues of Eq. (5.23) have to be real. The real eigenvalues and the associated eigenfunctions of Eq. (5.23) define a complete set of orthogonal functions (see Appendix D). With the help of these eigenfunctions any function can be reconstructed. Because the matrix operator $\underline{L}$ is neither positive nor negative definite, one will find both positive and negative eigenvalues $\left(\lambda_j^+, \lambda_j^-\right)$. The eigenfunctions belonging to these eigenvalues, $\left(\Phi_j^+, \Phi_j^-\right)$, form an orthogonal basis in the Hilbert space H. It is interesting to note that Eq. (5.23) reduces to the eigenvalue problem for the parabolic case, Eq. (3.25), if the Peclet number tends to infinity. Because the two sets of eigenvectors, normalized according to equation (5.25), constitute an orthogonal basis in H, an arbitrary vector $\vec{f}$ can be expanded in terms of eigenfunctions in the following way

$$\vec{f} = \sum_{j=0}^{\infty} D_j \, \vec{\Phi}_j(\tilde{n}) \tag{5.26}$$

The constants D_j in this equation can be calculated if we apply to both sides of Eq. (5.26) the inner product according to Eq. (5.14). This results in

$$\left\langle \vec{f}, \vec{\Phi}_i \right\rangle = \sum_{j=0}^{\infty} D_j \left\langle \vec{\Phi}_j, \vec{\Phi}_i \right\rangle \tag{5.27}$$

Because the expression $\left\langle \vec{\Phi}_j, \vec{\Phi}_i \right\rangle$ is equal to zero for $i \neq j$ (see Appendix D), Eq. (5.27) can be resolved for the constants D_j and one obtains

$$D_j = \frac{\left\langle \vec{f}, \vec{\Phi}_j \right\rangle}{\left\langle \vec{\Phi}_j, \vec{\Phi}_j \right\rangle} = \frac{\left\langle \vec{f}, \vec{\Phi}_j \right\rangle}{\left\| \vec{\Phi}_j \right\|^2} \tag{5.28}$$

where the vector norm $\left\| \vec{\Phi}_j \right\|^2$ has been introduced into Eq. (5.28). Inserting Eq. (5.28) into Eq. (5.27) results in the following expression for the expansion of an arbitrary vector $\vec{f}$ in terms of eigenfunctions

$$\vec{f} = \sum_{j=0}^{\infty} \frac{\left\langle \vec{f}, \vec{\Phi}_j \right\rangle}{\left\| \vec{\Phi}_j \right\|^2} \vec{\Phi}_j(\tilde{n}) \tag{5.29}$$

If we explicitly distinguish in equation (5.29) between positive and negative eigenvectors, the equation takes the following form

$$\vec{f} = \sum_{j=0}^{\infty} \frac{\left\langle \vec{f}, \vec{\Phi}_j^+ \right\rangle}{\left\| \vec{\Phi}_j^+ \right\|^2} \vec{\Phi}_j^+(\tilde{n}) + \sum_{j=0}^{\infty} \frac{\left\langle \vec{f}, \vec{\Phi}_j^- \right\rangle}{\left\| \vec{\Phi}_j^- \right\|^2} \vec{\Phi}_j^-(\tilde{n}) \tag{5.30}$$

After the expansion of an arbitrary vector in terms of eigenfunctions has been developed, we can return to the solution of the system of partial differential equations given by Eq. (5.10). Because of the non-homogeneous boundary condition for the temperature, the vector $\vec{S}$ does not belong to the domain $D(\underline{L})$ given by Eq. (5.15). However, it can be shown that

$$\left\langle \underline{L}\vec{S}, \vec{\Phi}_j \right\rangle = \left\langle \vec{S}, \underline{L}\vec{\Phi}_j \right\rangle + \Phi_{j2}(1) g(\tilde{x}) \tag{5.31}$$

holds (see Appendix D). The function $g(\tilde{x})$ is for the case of a semi-infinite heating zone given by

$$g(\tilde{x}) = \begin{cases} 1, & \tilde{x} \leq 0 \\ 0, & \tilde{x} > 0 \end{cases} \tag{5.32}$$

If we apply the inner product, given by Eq. (5.14), to the system of partial differential equations, Eq. (5.10), one obtains

$$\frac{\partial}{\partial \tilde{x}} \left\langle \vec{S}, \vec{\Phi}_j \right\rangle = \left\langle \underline{L}\vec{S}, \vec{\Phi}_j \right\rangle \tag{5.33}$$

Inserting Eq. (5.31) into Eq. (5.33) (and taking Eq. (5.20) into account) results in

$$\frac{\partial}{\partial \tilde{x}} \left\langle \vec{S}, \vec{\Phi}_j \right\rangle = \lambda_j \left\langle \vec{S}, \vec{\Phi}_j \right\rangle + g(\tilde{x}) \Phi_{j2}(1) \tag{5.34}$$

Eq. (5.34) represents an ordinary differential equation for the expression $\left\langle \vec{S}, \vec{\Phi}_j \right\rangle$. For the homogenous equation one obtains the solution

$$\left\langle \vec{S}, \vec{\Phi}_j \right\rangle = C_{0j} \exp(\lambda_j \tilde{x}) \tag{5.35}$$

A particular solution of Eq. (5.34) can be found by using the method of the variation of constants (see e.g. Bronstein and Semendjajew (1983)). One obtains after splitting the solution into positive and negative eigenvalues

$$\left\langle \vec{S},\vec{\Phi}_j^- \right\rangle = C_{0j}^- \exp(\lambda_j^- \tilde{x}) + \int_{-\infty}^{\tilde{x}} (g(\tilde{x})\,\Phi_{j2}^-(1))\exp(\lambda_j^- (\tilde{x}-\overline{x}))\,d\overline{x} \qquad (5.36)$$

$$\left\langle \vec{S},\vec{\Phi}_j^+ \right\rangle = C_{0j}^+ \exp(\lambda_j^+ \tilde{x}) - \int_{\tilde{x}}^{\infty} (g(\tilde{x})\,\Phi_{j2}^+(1))\exp(\lambda_j^+ (\tilde{x}-\overline{x}))\,d\overline{x} \qquad (5.37)$$

Because the solution must be bounded for $\tilde{x}\to+\infty$ and for $\tilde{x}\to-\infty$, the two constants C_{0j}^- and C_{0j}^+, appearing in Eqs. (5.36-5.37) have to be zero. The integral in Eqs. (5.36-5.37) can be evaluated by inserting the function $g(\tilde{x})$ into these equations. One obtains

$$\tilde{x}\le 0:\left\langle \vec{S},\vec{\Phi}_j \right\rangle = \frac{\Phi_{j2}^+(1)}{\lambda_j^+}\exp\left(\lambda_j^+ \tilde{x}\right) - \frac{\Phi_{j2}^+(1)}{\lambda_j^+} - \frac{\Phi_{j2}^-(1)}{\lambda_j^-} \qquad (5.38)$$

$$\tilde{x}> 0:\left\langle \vec{S},\vec{\Phi}_j \right\rangle = \frac{\Phi_{j2}^-(1)}{\lambda_j^-}\exp\left(\lambda_j^- \tilde{x}\right) \qquad (5.39)$$

Introducing these expressions into the expansion for the function $\vec{S}$, Eq. (5.29), and taking the first vector component of $\vec{S}$, the following result is obtained for the temperature distribution in the fluid

$$\tilde{x}\le 0:\Theta(\tilde{x},\tilde{n}) = -\sum_{j=0}^{\infty} \frac{\Phi_{j2}^-(1)\Phi_{j1}^-(\tilde{n})}{\lambda_j^- \left\|\vec{\Phi}_j^-\right\|^2} - \sum_{j=0}^{\infty} \frac{\Phi_{j2}^+(1)\Phi_{j1}^+(\tilde{n})}{\lambda_j^+ \left\|\vec{\Phi}_j^+\right\|^2} + \qquad (5.40)$$

$$\sum_{j=0}^{\infty} \frac{\Phi_{j2}^+(1)\Phi_{j1}^+(\tilde{n})}{\lambda_j^+ \left\|\vec{\Phi}_j^+\right\|^2}\exp(\lambda_j^+ \tilde{x})$$

$$\tilde{x}> 0:\Theta(\tilde{x},\tilde{n}) = -\sum_{j=0}^{\infty} \frac{\Phi_{j2}^-(1)\Phi_{j1}^-(\tilde{n})}{\lambda_j^- \left\|\vec{\Phi}_j^-\right\|^2}\exp(\lambda_j^- \tilde{x}) \qquad (5.41)$$

From Eqs (5.40–5.41) it can be seen that the solution for $\tilde{x}\le 0$ contains both, negative and positive eigenfunctions, whereas for $\tilde{x}>0$ only the negative eigenfunctions are needed for the solution of the problem. For Eq. (5.40), it is possible to show that

$$-\sum_{j=0}^{\infty} \frac{\Phi_{j2}^-(1)\Phi_{j1}^-(\tilde{n})}{\lambda_j^- \left\|\vec{\Phi}_j^-\right\|^2} - \sum_{j=0}^{\infty} \frac{\Phi_{j2}^+(1)\Phi_{j1}^+(\tilde{n})}{\lambda_j^+ \left\|\vec{\Phi}_j^+\right\|^2} = 1 \qquad (5.42)$$

This can be shown by expanding the vector $(1,0)^{\mathrm{T}}$ into a series according to Eq. (5.29) (see Appendix D for a derivation of this result). Introducing this result into Eq. (5.40) leads to

$$\tilde{x} \leq 0 : \Theta(\tilde{x}, \tilde{n}) = 1 + \sum_{j=0}^{\infty} \frac{\Phi_{j2}^{+}(1)\Phi_{j1}^{+}(\tilde{n})}{\lambda_j^{+} \left\| \vec{\Phi}_j^{+} \right\|^2} \exp(\lambda_j^{+} \tilde{x}) \tag{5.43}$$

From Eqs. (5.41) and (5.43) one can see that they result in a continuous temperature distribution for $x = 0$. In addition, one notices from the Eqs. (5.41) and (5.43) that these equations satisfy all the boundary conditions given by Eq. (5.7).

If the Peclet number increases, the eigenvalues λ_j^{+} tend to infinity and Eq. (5.43) (the temperature distribution for $\tilde{x} \leq 0$) results in

$$\Theta(\tilde{x}, \tilde{n}) = 1, \quad \text{for Pe}_L \to \infty \tag{5.44}$$

which represents the solution for the parabolic case. Eq. (5.44) also elucidates, that it is only correct to prescribe a temperature boundary condition for $\tilde{x} = 0$ for the limiting case of $\text{Pe}_L \to \infty$.

The vector norm $\left\| \vec{\Phi}_j \right\|^2$ which has been used in the above given solution for the temperature field, can be rewritten by using the Eqs. (5.21-5.22). One obtains (see Appendix D)

$$\left\| \vec{\Phi}_j \right\|^2 = -\frac{1}{\lambda_j} \int_0^1 \tilde{r}^F \tilde{u} \, \Phi_{j1}^2 d\tilde{n} + \frac{2}{\text{Pe}_L^2} \int_0^1 \tilde{r}^F a_1(\tilde{n}) \Phi_{j1}^2 d\tilde{n} \tag{5.45}$$

Eq. (5.45) shows an interesting result: the second term in this equation depends on the Peclet number and tends to zero for increasing values of Pe_L. This means that, for large Peclet numbers, the second term in Eq. (5.45) can be neglected and therefore, the temperature distribution, given by Eq. (5.41), is formally identical to the distribution for the parabolic problem. In addition, it can be shown from the eigenvalue problem, Eq. (5.23), that this equation reduces to one for the parabolic problem for $\text{Pe}_L \to \infty$. This shows nicely that the results presented here approach those given in Chap. 3 for large Peclet numbers. Furthermore, it can be shown that

$$\left\| \vec{\Phi}_j \right\|^2 = \Phi_{j2}(1) \frac{d\Phi_{j1}(1)}{d\lambda} \bigg|_{\lambda_j} \tag{5.46}$$

(see Appendix D). If one introduces the abbreviation

$$A_j = \left(\lambda_j \frac{d\Phi_{j1}(1)}{d\lambda} \bigg|_{\lambda = \lambda_j} \right)^{-1} \tag{5.47}$$

into the equation above, one finally obtains for the temperature field in the flow

$$\tilde{x} \leq 0: \; \Theta(\tilde{x}, \tilde{n}) = 1 + \sum_{j=0}^{\infty} A_j^+ \, \Phi_{j1}^+(\tilde{n}) \exp(\lambda_j^+ \, \tilde{x}) \tag{5.48}$$

$$\tilde{x} > 0: \; \Theta(\tilde{x}, \tilde{n}) = \sum_{j=0}^{\infty} A_j^- \, \Phi_{j1}^-(\tilde{n}) \exp(\lambda_j^- \, \tilde{x}) \tag{5.49}$$

After the temperature field in the fluid is known, the bulk-temperature and the Nusselt number can be calculated. Introducing Eqs. (5.48-5.49) into the definition of the bulk-temperature, Eq. (3.67), and of the Nusselt number, Eq. (3.66), results in

$$\tilde{x} \leq 0: \; \Theta_b = 1 + 2^F \sum_{j=0}^{\infty} A_j^+ \left\{ \frac{\Phi_{j1}'^+(1)}{\lambda_j^+} + \frac{\lambda_j^+}{Pe_L^2} \int_0^1 a_1(\tilde{n}) \Phi_{j1}^+(\tilde{n}) \tilde{r}^F \, d\tilde{n} \right\} \exp(\lambda_j^+ \, \tilde{x}) \tag{5.50}$$

$$\tilde{x} > 0: \; \Theta_b = 1 + 2^F \sum_{j=0}^{\infty} A_j^- \left\{ \frac{\Phi_{j1}'^-(1)}{\lambda_j^-} + \frac{\lambda_j^-}{Pe_L^2} \int_0^1 a_1(\tilde{n}) \Phi_{j1}^-(\tilde{n}) \tilde{r}^F \, d\tilde{n} \right\} \exp(\lambda_j^- \, \tilde{x}) \tag{5.51}$$

$$\tilde{x} \leq 0: \; Nu_D = \frac{-4 \sum_{j=0}^{\infty} A_j^+ \, \Phi_{j1}'^+(1) \exp(\lambda_j^+ \, \tilde{x})}{4^F \sum_{j=0}^{\infty} A_j^+ \left\{ \frac{\Phi_{j1}'^+(1)}{\lambda_j^+} + \frac{\lambda_j^+}{Pe_L^2} \int_0^1 a_1(\tilde{n}) \Phi_{j1}^+(\tilde{n}) \tilde{r}^F \, d\tilde{n} \right\} \exp(\lambda_j^+ \, \tilde{x})} \tag{5.52}$$

$$\tilde{x} > 0: \; Nu_D = \frac{-4 \sum_{j=1}^{\infty} A_j^- \, \Phi_{j1}'^-(1) \exp(\lambda_j^- \, \tilde{x})}{4^F \sum_{j=1}^{\infty} A_j^- \left\{ \frac{\Phi_{j1}'^-(1)}{\lambda_j^-} + \frac{\lambda_j^-}{Pe_L^2} \int_0^1 a_1(\tilde{n}) \Phi_{j1}^-(\tilde{n}) \tilde{r}^F \, d\tilde{n} \right\} \exp(\lambda_j^- \, \tilde{x})} \tag{5.53}$$

where $F = 0$ denotes the heat transfer in a parallel plate channel, whereas $F = 1$ denotes the heat transfer in a pipe. From Eqs. (5.52-5.53) the Nusselt number for the fully developed flow can be obtained. Investigating the limiting case $\tilde{x} \to \infty$ in Eq. (5.53), only the first term of the sum needs to be retained and one obtains

$$Nu_\infty^- = \frac{4}{4^F} \left| \lambda_0^- \right| \Big/ \left\{ 1 + \frac{\lambda_0^{-2}}{\Phi_{01}'^-(1) Pe_L^2} \int_0^1 a_1(\tilde{n}) \Phi_{01}^-(\tilde{n}) \tilde{r}^F \, d\tilde{n} \right\} \tag{5.54}$$

This equation shows again, that for $Pe_L \to \infty$ the Nusselt number for the fully developed flow is given by an expression similar to Eq. (3.70). For finite values of the Peclet number, this value is changed by the second term in the denominator of Eq. (5.54).

5.1.1 Heat Transfer in Laminar Pipe and Channel Flows for Small Peclet Numbers

For laminar flows the functions a_1 and a_2 are equal to one. In addition, the velocity distribution is given by a simple parabolic expression, according to Eqs. (3.1-3.2). As stated before, the eigenvalues and eigenfunctions of Eq. (5.23) have to be calculated numerically. This can be done with the method described in Appendix C.

Laminar Pipe Flow

In order to investigate this case, the flow index F in all the equations has to be set to 1. This case has been investigated with the method of Papoutsakis et al. (1980a). Table 5.1 shows the first fifteen eigenvalues and constants for two different Peclet numbers. It can be seen that the positive eigenvalues strongly increase with growing values of the Peclet number. For example, the first positive eigenvalue (λ_0^+) increases by about a factor of three by increasing the Peclet number from 5 to 10. Since larger eigenvalues result, according to Eq. (5.48), in a faster decrease of the influence of axial heat conduction, this elucidates how this effect diminishes with increasing values of the Peclet number. The increase of the positive eigenvalues with growing Peclet numbers can be seen clearly in Fig. 5.2, where the first 100 positive eigenvalues are plotted for four different Peclet numbers.

Table 5.1: Eigenvalues and constants for laminar pipe flow for $Pe_D = 5$ and $Pe_D = 10$ (adapted from Papoutsakis et al. (1980a) and own calculations).

	$Pe_D = 5$		$Pe_D = 10$	
j	λ_j^+	A_j^+	λ_j^+	A_j^+
0	12.289076	-0.342609	37.383579	-0.107926
1	18.929129	0.412438	51.950323	0.232838
2	26.360733	-0.374175	64.468355	-0.303513
3	34.041020	0.334422	78.791847	0.296627
4	41.799450	-0.303680	93.763713	-0.279034
5	49.593035	0.279668	109.009770	0.262030
6	57.405409	-0.260405	124.400395	-0.247011
7	65.229022	0.244561	139.876976	0.233958
8	73.059822	-0.231254	155.408680	-0.222598
9	80.895593	0.219881	170.978041	0.212646
10	88.734839	-0.210023	186.574029	-0.203861
11	96.576707	0.201373	202.189809	0.196042
12	104.420491	-0.193702	217.820419	-0.189034
13	112.265848	0.186843	233.462654	0.182711
14	120.112354	-0.180663	249.114013	-0.176971

j	$-\lambda_j^-$	A_j^-	$-\lambda_j^-$	A_j^-
0	2.844835	1.233275	3.372024	1.379949
1	10.174296	-0.617349	15.383971	-0.680027
2	17.859425	0.464093	29.751723	0.493104
3	25.617868	-0.388494	44.738343	-0.406545
4	33.410361	0.340928	59.986307	0.354040
5	41.221752	-0.307382	75.375199	-0.317634
6	49.044572	0.282080	90.849326	0.290432
7	56.874853	-0.262122	106.378881	-0.269111
8	64.710161	0.245860	121.946337	0.251826
9	72.549123	-0.232280	137.540929	-0.237452
10	80.390694	0.220720	153.155416	0.225259
11	88.234332	-0.210726	168.785154	-0.214752
12	96.079491	0.201973	184.426542	0.205575
13	103.925919	-0.194223	200.077324	-0.197472
14	111.773292	0.187302	215.735491	0.190250

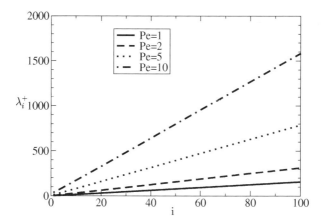

Fig. 5.2: Distribution of the positive eigenvalues for different Peclet numbers

Fig. 5.3 shows the temperature distribution along the pipe centerline for different Peclet numbers. It can be seen how the axial heat conduction effect in the fluid decreases rapidly with increasing values of the Peclet number. In addition, it can be seen from this figure that, for a Peclet number of 5, the region of influence of the axial heat conduction is about one radius in the negative direction.

Fig. 5.4 shows the development of the temperature profile for different axial positions in the pipe (Papoutsakis et al. (1980a)). Note that, for growing values of the axial coordinate, the temperature tends to T_W and thus Θ tends to zero. For $\tilde{x} \leq 0$ the temperature of the fluid rapidly decreases to T_0, resulting in values of Θ close to one for increasing distances from $\tilde{x} = 0$. By comparing the two plots in Fig. 5.4, one can notice how the influence of the axial heat conduction in the flow

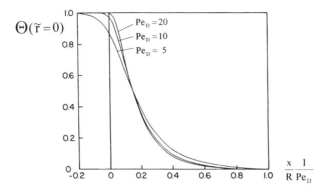

Fig. 5.3: Temperature distribution at the pipe centerline for different Peclet numbers (Papoutsakis, Ramkrishna and Lim (1980a))

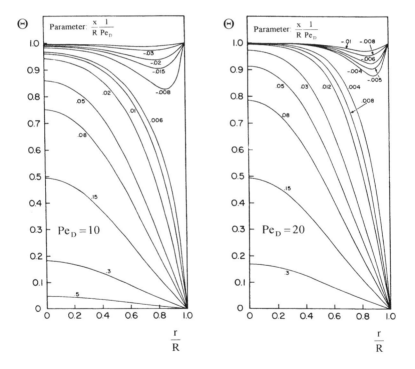

Fig. 5.4: Temperature distribution for several axial positions $x/(R\,Pe_D)$ and two Peclet numbers (Papoutsakis, Ramkrishna and Lim (1980a))

gets weaker with increasing values of the Peclet number. For $Pe_D = 20$, the temperature distribution at the entrance of the heating section ($\tilde{x} = 0$) becomes rather non-uniform. This shows clearly that whenever the axial heat conduction is not negligible, it is incorrect to prescribe a constant entrance temperature at $\tilde{x} = 0$. In this case, the temperature distribution at $\tilde{x} = 0$ is always a result of the solution. Papoutsakis et al. (1980a) calculated the value of the Nusselt number for fully developed flow Nu_∞^-, according to Eq. (5.54), and showed graphically the distribution as function of the Peclet number. In general, it can be seen that Nu_∞^- increases with decreasing values of the Peclet number. Hennecke (1968) predicted numerically the value of Nu_∞^- for several Peclet number. These values have been tabulated by Shah and London (1978). Table 5.2 shows a comparison between analytically and numerically predicted values of Nu_∞^-. From Table 5.2, a very good agreement between numerically and analytically predicted values can be noted.

Table 5.2: Comparison between numerically (Hennecke (1968), values taken from Shah and London (1978)) and analytically (values in the Table have been predicted by the author using the equations given by Papoutsakis et al. (1980a)) predicted values for Nu_∞^-

Pe_D	Nu_∞^- (Papoutsakis et al. (1980a))	Nu_∞^- (Hennecke (1968))
1	4.027	4.03
2	3.922	3.92
5	3.767	3.77
10	3.695	3.70
20	3.668	3.67
50	3.659	3.66

In Figs. 5.5–5.6, the distribution of the local Nusselt number is shown for laminar pipe flow for $\tilde{x} > 0$. These values have been calculated using the numerical approach described in detail in Appendix C. In order to resolve the distribution of the Nusselt number for very small values of $\tilde{x}$ correctly, a very large number of eigenvalues is needed. For the present calculations 200 eigenvalues have been taken. If less eigenvalues are used, the Nusselt number for $\tilde{x} \rightarrow 0$ will tend to smaller values. This can be seen clearly, if the Nusselt number is plotted on a logarithmic scale. In addition to the large number of eigenvalues, a very fine grid was used in order to resolve the shape of the higher eigenfunctions (see Appendix C). Normally 1500 – 2000 grid points in the radial direction are sufficient. As it can be seen from the Figs. 5.5-5.6, the analytical method is in excellent agreement with the numerical calculations by Hennecke (1968) for the two shown Peclet numbers. For $Pe_D = 50$, the analytical solution has also been compared with the investigations of Bayazitoglu and Özisik (1980) and of Singh (1958) and a very good agreement was found.

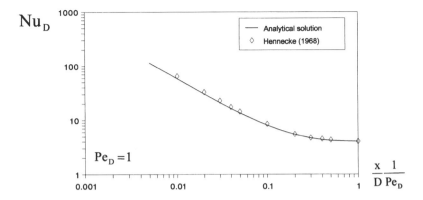

Fig. 5.5: Distribution of the local Nusselt number for $\tilde{x} > 0$ and $Pe_D = 1$

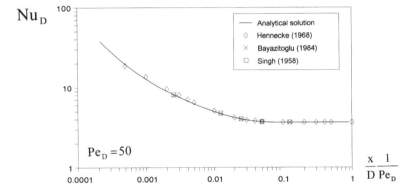

Fig. 5.6: Distribution of the local Nusselt number for $\tilde{x} > 0$ and $Pe_D = 50$

Laminar Channel Flow

In order to investigate this case, the flow index F in all the equations has to be set to 0. The first solution to this problem was obtained by Deavours (1974). He decomposed the eigenvalue problem, for the parallel plate channel, into a system of ordinary differential equations for which he proved the orthogonality of the eigenfunctions. As stated before, this case was also investigated by Weigand et al. (1993) with the analytical method presented here. Table 5.3 shows eigenvalues and constants for two different Peclet numbers. Similar to the circular pipe case (Table 5.1), it can be seen from Table 5.3 that the positive eigenvalues increase dramatically with growing values of the Peclet number. This shows again that the effect of the axial heat conduction in the fluid diminishes with growing values of the Peclet number.The value of the Nusselt number for fully developed channel flow, Nu_∞^-, has also been reported in literature. Table 5.4 shows a comparison between the analytical predicted values and calculations of Ash (see Shah and London (1978)) and Nguyen (1992). The values are ranging from $\mathrm{Pe}_D = 0.02352$ up to $\mathrm{Pe}_D = 1000$ and show a very good agreement between the presented analytical solution and literature data. It can be seen that the Nusselt number for the fully developed flow increases with decreasing values of the Peclet number. This behaviour is very similar to the results shown in the previous section for the circular pipe.

Table 5.3: Eigenvalues and constants for laminar channel flow ($\mathrm{Pe}_D = 5$ and $\mathrm{Pe}_D = 10$)

	$\mathrm{Pe}_D = 5$		$\mathrm{Pe}_D = 10$	
j	λ_j^+	A_j^+	λ_j^+	A_j^+
0	3.220469	-0.376403	9.575077	-0.230822
1	6.758122	0.202738	15.522013	0.192715
2	10.645334	-0.124312	23.101943	-0.121091
3	14.557251	0.089495	30.850355	0.087928
4	18.476438	-0.069885	38.648148	-0.068971
5	22.398690	0.057315	46.467380	0.056719
6	26.322503	-0.048575	54.297781	-0.048157
7	30.247224	0.042147	62.134735	0.041837
8	34.172509	-0.037220	69.975845	-0.036981
9	38.098175	0.033325	77.819763	0.033135
10	42.024105	-0.030167	85.665658	-0.030013
11	45.950232	0.027556	93.513006	0.027428
12	49.876499	-0.025361	101.361442	-0.025253
13	53.802882	0.023489	109.210726	0.023398
14	57.729349	-0.021875	117.060670	-0.021796

j	$-\lambda_j^-$	A_j^-	$-\lambda_j^-$	A_j^-
0	1.192001	0.897756	1.580670	1.044165
1	5.139142	-0.222463	8.989775	-0.232941
2	9.062116	0.130123	16.748478	0.132737
3	12.984172	-0.092272	24.547406	-0.093479
4	16.907540	0.071518	32.366159	0.072234
5	20.831909	-0.058393	40.195993	-0.058873
6	24.756940	0.049341	48.032482	0.049686
7	28.682423	-0.042719	55.873251	-0.042979
8	32.608218	0.037664	63.716909	0.037868
9	36.534242	-0.033679	71.562614	-0.033843
10	40.460433	0.030456	79.409807	0.030592
11	44.386753	-0.027797	87.258131	-0.027910
12	48.313173	0.025564	95.107316	0.025661
13	52.239673	-0.023664	102.957190	-0.023747
14	56.166236	0.022026	110.807606	0.022098

Table 5.4: Nusselt number for the fully developed flow , Nu_∞^-, compared to predictions by Ash (see Shah and London (1978)) and Nguyen (1992).

Pe_D	Nu_∞^- (Eq. (5.54))	Nu_∞^- Ash (see Shah and London (1978))	Nu_∞^- Nguyen (1992)
0.02352	8.1141	8.1144	
0.4444	8.0644	8.0644	
0.6508	8.0416	8.0416	
1.0576	7.9997	7.9998	
1.4368	7.9640	7.9640	
2	7.9165		7.9164
5	7.7471		7.7468
9.97	7.6310	7.6310	
50	7.5457		7.5456
69.78	7.5432	7.5408	
100	7.5419		7.5407
1000	7.5407		7.5407

The temperature distribution at the centreline of the fluid is depicted in Fig. 5.7. By comparing Fig. 5.7 with Fig. 5.3 for the circular pipe, it can be seen that the axial heat conduction effect within the fluid has very similar effects for the two geometries. In addition, it can be seen that for very low values of the Peclet number the temperature at $\tilde{x} = 0$ is lowered because of axial heat conduction effects within the flow.

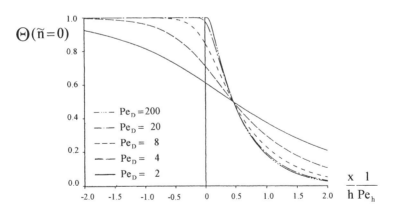

Fig. 5.7: Temperature distribution at the centerline of the parallel plate channel for different Peclet numbers

Fig. 5.8 shows the distribution of the local Nusselt number for $Pe_D = 4$ and $\tilde{x} > 0$. The Nusselt number is compared to calculated values of Agrawal (1960). Agrawal (1960) determined for $\tilde{x} > 0$ and $\tilde{x} \le 0$ two independent series solutions, which then have been matched at $\tilde{x} = 0$. This approach resulted in a complicated expression for the Nusselt number. From Fig. 5.8, it can be seen that both calculations are in good agreement for $\tilde{x} \geq 0.1$. However, for smaller values of $\tilde{x}$, the values given by Agrawal (1960) are much lower than the Nusselt numbers predicted here. The reason for this behavior is caused by the fact that Agrawal used only five eigenvalues for the prediction of the local Nusselt number. This is sufficient for larger values of the axial coordinate, but not for very small values of the axial coordinate[2]. For very small values of the axial coordinate, it can be seen from Fig. 5.8 that the Nusselt number can be approximated by $\mathrm{Nu}_D \propto \tilde{x}^{-|\beta|}$. This is in good agreement with predictions by Hennecke (1968) and by Kader (1971).

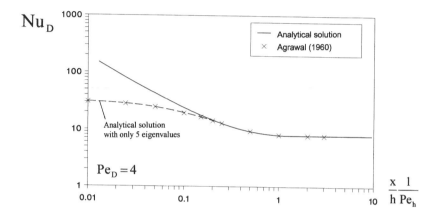

Fig. 5.8: Distribution of the local Nusselt number for $\mathrm{Pe}_D = 4$ $(\tilde{x} > 0)$

5.1.2 Heat Transfer in Turbulent Pipe and Channel Flows for Small Peclet Numbers

For the heat transfer in a turbulent duct flow, the functions a_1 and a_2 depend on the turbulent heat exchange in the flow. For the following analysis, the ratios $\varepsilon_{hx} / \varepsilon_{hr}$ and $\varepsilon_{hx} / \varepsilon_{hy}$ are set equal to one. This assumption has been made previously by Lee (1982) and Chieng and Launder (1980). Weigand et al. (2002) proved this assumption by solving Eq. (3.6) numerically by using a non-isotropic heat flux model for the heat transfer in a planar channel. They concluded that the above assumption is well justified for the applications considered here. Assuming that the

---

[2] If only the first five eigenvalues for the prediction of the Nusselt number are used with the present method, the predicted values are in close agreement with the ones by Agrawal (1960) (see Fig. 5.8).

ratio $\varepsilon_{hx}/\varepsilon_{hn}$ is equal to one, results in the fact that a_1 is equal to a_2 (see Eq. (5.8)). Since axial heat conduction effects in turbulent flows are only important if the Prandtl number is very small, the turbulent Prandtl number has to be calculated by adopting a model suitable for liquid metal flows (e.g. Eqs. (3.79-3.82)).

Turbulent Pipe Flow

If the flow index F is set to one in the equations, the turbulent heat transfer in a pipe for a low Peclet number flow can be predicted. For the following calculations the turbulent Prandtl number model of Azer and Chao (1960) (see Eq. (3.82)) as well as the Kays and Crawford model (1993) (see Eqs. (3.79-3.81)) have been used. These two models have been found to give reliable answers for turbulent liquid metal flows. The reader is referred to Weigand (1996) or to Weigand et al. (1997) for comparisons between calculations and measurements for liquid metal flows in pipes with the above mentioned models for Pr_t.

Table 5.5: Eigenvalues and constants for turbulent pipe flow with Pr = 0.001.

j	$\mathrm{Re}_D = 5000$		$\mathrm{Re}_D = 10000$	
	λ_j^+	A_j^+	λ_j^+	A_j^+
0	10.915	-0.3959	33.569	-0.2064
1	17.925	0.4220	45.293	0.3225
2	25.530	-0.3748	59.584	-0.3210
3	33.266	0.3354	74.598	0.3018
4	41.051	-0.3049	89.896	-0.2820
5	48.859	0.2810	105.33	0.2644
6	56.628	-0.2617	120.85	-0.2491
7	64.512	0.2458	136.41	0.2359
8	72.348	-0.2324	152.01	-0.2244
9	80.189	0.2209	167.63	0.2144

j	$-\lambda_j^-$	A_j^-	$-\lambda_j^-$	A_j^-
0	3.305	1.2055	4.280	1.3923
1	10.626	-0.6417	16.837	-0.7376
2	18.337	0.4758	31.444	0.5281
3	26.127	-0.3936	46.622	-0.4262
4	33.947	0.3430	62.023	0.3653
5	41.779	-0.3080	77.532	-0.3242
6	49.617	0.2820	93.102	0.2943
7	57.460	-0.2618	108.71	-0.2713
8	65.305	0.2453	124.34	0.2530
9	73.152	-0.2317	139.98	0.2380

Eigenvalues and constants are shown in Table 5.5 for a Prandtl number of 0.001 and different values of the Reynolds number. It can be seen from Table 5.5 that the positive eigenvalues increase rapidly with growing values of the Reynolds number. This shows again the decreasing influence of the axial heat conduction within the flow.

Fig. 5.9 shows the distribution of the Nusselt number for fully developed flow (Nu_∞^-) as a function of the Peclet number for Pr = 0.022. The predicted values have been obtained by using different models for the turbulent Prandtl number. It can be seen that the two models give relatively similar results. The predicted values have been compared to experimental data from Sleicher et al. (1973) and from Gilliland et al. (1951) (taken from Azer and Chao (1960)). As it can be seen from Fig. 5.9, the agreement between experimental data and predictions is good.

Table 5.6 shows values of the Nusselt number Nu_∞^- for fully developed flow for different values of the Prandtl and Reynolds number. In addition, the table contains the relative deviation of the Nusselt number for a parabolic calculation $\Delta E = \left(Nu_{elliptic} - Nu_{parabolic} \right) / Nu_{elliptic}$.

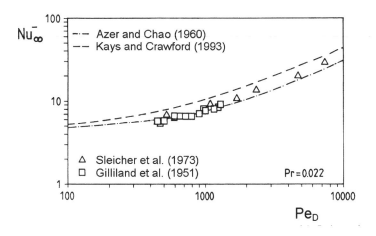

Fig. 5.9: Nusselt number of the fully developed flow (Nu_∞^-) as a function of the Peclet number for Pr = 0.022 (Weigand, Ferguson and Crawford (1997))

From the entries in Table 5.6, it is obvious that axial heat conduction effects are not important for the fully developed flow. Therefore, the influence of axial heat conduction on the Nusselt number for fully developed flow can be neglected with good accuracy. This may not be the case when the thermal entrance region is considered.

Fig. 5.10 shows the relative error ΔE in the local Nusselt number by ignoring the axial heat conduction within the thermal entrance region. As mentioned before, the relative error is negligible for the fully developed flow ($\tilde{x} \to \infty$). However, it may be very large for small values of the axial coordinate. For example for $Pe_D = 10$, the error will be about 30% for $x/(D\,Pe_D) = 0.01$ and for smaller values of the axial coordinate this error is even larger.

Table 5.6: Nusselt number for fully developed flow (Nu_∞^-). The values in brackets indicate the relative deviation $\Delta E = \left(Nu_{elliptic} - Nu_{parabolic} \right) / Nu_{elliptic}$ between an elliptic and a parabolic calculation.

Pr	0.002	0.006	0.01
Re_D			
3000	8.370	8.284	8.277
	(0.014)	(0.003)	(0.001)
5000	8.518	8.484	8.497
	(0.007)	(0.001)	(0.0004)
10000	8.760	8.783	8.835
	(0.002)	(0.0002)	(0.0001)
20000	8.919	9.011	9.149
	(0.0004)	(0.0)	(0.0)
30000	9.012	9.175	9.408
	(0.0002)	(0.0)	(0.0)

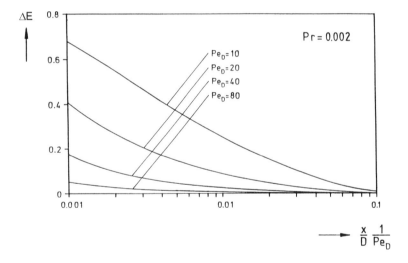

Fig. 5.10: Relative error in the local Nusselt number by neglecting the axial heat conduction within the fluid (Weigand (1996))

This example shows very clearly that axial heat conduction effects may be important in the thermal entrance region, also for turbulent flows. Fig. 5.11 shows the local temperature field in the flow for two different values of the Peclet number. The increasing effect of the axial heat conduction in the flow for decreasing values of the Peclet number is clearly visible for small axial values.

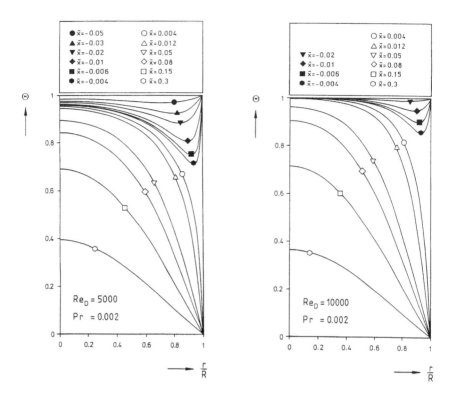

Fig. 5.11: Temperature distribution for a thermally developing flow in a circular pipe (Weigand (1996))

Turbulent Flow in a Planar Channel

For flow and heat transfer in a parallel plate channel, the flow index F in the pre-ceeding equations must be set to 0. The behaviour of the solutions, e.g. for the Nusselt number, is very similar to the one for the circular pipe and therefore is not been presented here. The reader is referred to Weigand (1996) and Weigand et al. (2002) for a detailed presentation of these results.

5.2 Heat Transfer for Constant Wall Heat Flux for $x \leq 0$ and $x > 0$

This case has been studied analytically by Papoutsakis et al. (1980b) for laminar, hydrodynamically fully developed pipe flow and by Weigand et al. (2001) for laminar and turbulent hydrodynamically fully developed pipe and channel flows.

The geometry under consideration and the used coordinate system are shown in Fig. 5.12

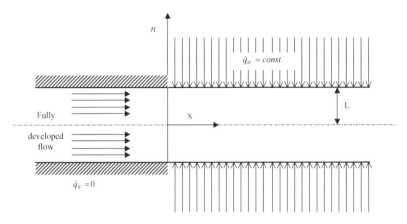

Fig. 5.12: Geometrical configuration and coordinate system

Similar to the considerations given in Chap. 5.1, we start with Eq. (5.3)

$$\rho c_p\, u(n)\, \frac{\partial T}{\partial x} = \frac{1}{r^F} \frac{\partial}{\partial n}\left[r^F \left(k + \rho c_p \frac{\varepsilon_m}{\mathrm{Pr}_t} \right) \frac{\partial T}{\partial n} \right] + \frac{\partial}{\partial x}\left[\left(k + \rho c_p \frac{\varepsilon_m}{\mathrm{Pr}_t} \right) \frac{\partial T}{\partial x} \right] \tag{5.3}$$

This equation has to be solved with the boundary conditions

$$n = 0 : \partial T / \partial n = 0, \tag{5.55}$$
$$n = L : x \leq 0 : \partial T / \partial n = 0,$$
$$x > 0 : \partial T / \partial n = \dot{q}_w / k$$
$$\lim_{x \to -\infty} T = T_0$$

Introducing the following dimensionless quantities

$$\Theta = \frac{T - T_0}{\dot{q}_W L / k},\; \tilde{x} = \frac{x}{L} \frac{1}{\mathrm{Pe}_L},\; \tilde{u} = \frac{u}{\overline{u}},\; \tilde{n} = \frac{n}{L},\; \tilde{r} = \frac{r}{L} \tag{5.56}$$

$$\mathrm{Pe}_L = \mathrm{Re}_L \mathrm{Pr},\; \mathrm{Re}_L = \frac{\overline{u}\, L}{v},\; \mathrm{Pr} = \frac{v}{a},\; \tilde{\varepsilon}_m = \frac{\varepsilon_m}{v},\; \mathrm{Pr}_t = \frac{\varepsilon_m}{\varepsilon_{hn}}$$

into the energy equation (5.3) and into the boundary conditions, Eq. (5.55), results in

$$\tilde{u}(\tilde{n})\, \frac{\partial \Theta}{\partial \tilde{x}} = \frac{1}{\mathrm{Pe}_L^2} \frac{\partial}{\partial \tilde{x}}\left[a_1(\tilde{n}) \frac{\partial \Theta}{\partial \tilde{x}} \right] + \frac{1}{\tilde{r}^F} \frac{\partial}{\partial \tilde{n}}\left[\tilde{r}^F a_2(\tilde{n}) \frac{\partial \Theta}{\partial \tilde{n}} \right] \tag{5.6}$$

with the boundary conditions

$$\tilde{x} \to -\infty : \Theta = 0$$

$$\tilde{n} = 0: \frac{\partial \Theta}{\partial \tilde{n}} = 0$$

$$\tilde{n} = 1: \ \tilde{x} > 0: \frac{\partial \Theta}{\partial \tilde{n}} = 1, \qquad \tilde{x} \le 0: \frac{\partial \Theta}{\partial \tilde{n}} = 0$$

(5.57)

Similar to the procedure shown in Chap. 3.3, the temperature for the present case has been made dimensionless with the constant wall heat flux and the thermal conductivity of the fluid. The functions a_1 and a_2 are given by Eq. (5.8). The solution procedure for the energy equation for a constant heat flux at the wall is similar to the one for constant wall temperature. We are introducing again the axial energy flow through a cross-sectional area of the height $\tilde{n}$ ($E(\tilde{x}, \tilde{n})$), given by Eq. (5.9). This results again in a system of partial differential equations, Eq. (5.10), with the solution vector $\vec{S}$ and the matrix operator $\underset{\sim}{L}$ defined by Eq. (5.11). The boundary conditions for the function $E(\tilde{x}, \tilde{n})$ can be derived from the definition of the axial energy flux, Eq. (5.9) by using Eq. (5.57). One obtains

$$\lim_{\tilde{x} \to -\infty} E = 0 \ , \qquad , \ \tilde{n} = 0 : E = 0$$

(5.58)

$$\tilde{n} = 1 : E = \begin{cases} 0, & -\infty < \tilde{x} \le 0 \\ \tilde{x}, & \tilde{x} > 0 \end{cases}$$

For the inner product, the expression given by Eq. (5.14) can be used again. However, because of the changed boundary conditions, $D(\underset{\sim}{L})$ needs to be adapted

$$D(\underset{\sim}{L}) = \left\{ \vec{\Phi} \in H : \underset{\sim}{L} \vec{\Phi} \text{(exists and)} \in H, \Phi_2 (1) = \Phi_2 (0) = 0 \right\}$$

(5.59)

where the only change in Eq. (5.59) compared to Eq. (5.15) lies in the changed boundary condition that $\Phi_2 (1) = 0$.

For the eigenvalue problem one obtains again

$$\left[\tilde{r}^F a_2 (\tilde{n}) \Phi'_{j1} \right]' + \tilde{r}^F \left[\frac{\lambda_j a_1 (\tilde{n})}{Pe_L^2} - \tilde{u} \right] \lambda_j \Phi_{j1} = 0$$

(5.23)

but now with the boundary conditions

$$\Phi'_{j1}(0) = 0 \qquad , \Phi'_{j1}(1) = 0$$

(5.60)

Comparing the boundary conditions given by Eq. (5.60) for a constant wall heat flux to the ones for a constant wall temperature, it can be seen that only the boundary condition for $\tilde{n} = 1$ has been changed. The normalizing condition for the eigenfunctions, Eq. (5.25), can be used again. Similarly as for Eq. (5.26) it can be shown that an arbitrary vector $\vec{f}$ can be expanded into a series of eigenfunctions and the result is given by Eq. (5.30).

Since the expansion of an arbitrary vector in terms of eigenfunction has already been developed, we focus directly on the solution of the system of partial differential equations given by Eq. (5.10). Because of the non-homogeneous boundary condition for the heat flux, the vector $\vec{S}$ does not belong to the domain $D(\underset{\sim}{L})$ given by Eq. (5.59). However, it can be shown that

$$\left\langle \underset{\sim}{L}\vec{S}, \vec{\Phi}_j \right\rangle = \left\langle \vec{S}, \underset{\sim}{L}\vec{\Phi}_j \right\rangle - \Phi_{j1}(1)E(\tilde{x}, 1) \tag{5.61}$$

Note here the difference in this equation compared to Eq. (5.31). The last term on the right hand side of Eq. (5.61) is now a function of the axial coordinate and no longer a constant. Taking the inner product of both sides of Eq. (5.61) one obtains

$$\frac{\partial}{\partial \tilde{x}} \left\langle \vec{S}, \vec{\Phi}_j \right\rangle = \lambda_j \left\langle \vec{S}, \vec{\Phi}_j \right\rangle - \Phi_{j1}(1)E(\tilde{x}, 1) \tag{5.62}$$

Eq. (5.62) represents an ordinary differential equation for the expression $\left\langle \vec{S}, \vec{\Phi}_j \right\rangle$ and can be solved separately for positive and negative eigenvalues. This results in

$$\left\langle \vec{S}, \vec{\Phi}_j^- \right\rangle = -\Phi_{j1}^-(1) \int_{-\infty}^{\tilde{x}} E(\tilde{x}, 1)\exp(\lambda_j^-(\tilde{x} - \overline{x}))\,d\overline{x} \tag{5.63}$$

$$\left\langle \vec{S}, \vec{\Phi}_j^+ \right\rangle = \Phi_{j1}^+(1) \int_{-\infty}^{\tilde{x}} E(\tilde{x}, 1)\exp(\lambda_j^+(\tilde{x} - \overline{x}))\,d\overline{x} \tag{5.64}$$

After evaluating the integrals in Eqs. (5.63-5.64), the following result for the temperature distribution Θ, which is the first vector component of $\vec{S}$, can be derived

$$\tilde{x} \leq 0: \ \Theta(\tilde{x}, \tilde{n}) = \sum_{j=1}^{\infty} \frac{\Phi_{j1}^+(1)}{\left\| \vec{\Phi}_j^+ \right\|^2} \frac{\exp\left(\lambda_j^+ \tilde{x}\right)}{\lambda_j^{+2}} \Phi_{j1}^+(\tilde{n}) \tag{5.65}$$

$$\tilde{x} > 0: \ \Theta(\tilde{x}, \tilde{n}) = \sum_{j=0}^{\infty} \frac{\Phi_{j1}(1)}{\lambda_j^2 \left\| \vec{\Phi}_j \right\|^2} \Phi_{j1}(\tilde{n}) + \tilde{x} \sum_{j=0}^{\infty} \frac{\Phi_{j1}(1)}{\lambda_j \left\| \vec{\Phi}_j \right\|^2} \Phi_{j1}(\tilde{n})$$
$$- \sum_{j=0}^{\infty} \frac{\Phi_{j1}^-(1)\Phi_{j1}^-(\tilde{n})}{\lambda_j^{-2} \left\| \vec{\Phi}_j^- \right\|^2} \exp(\lambda_j^- \tilde{x}) \tag{5.66}$$

The temperature distribution given above contains for $\tilde{x} > 0$ both negative and positive eigenfunctions. The expressions for the temperature field can be further simplified by replacing the first two terms in Eq. (5.66). This results (see Appendix D) in

$$\tilde{x} \le 0: \quad \Theta(\tilde{x}, \tilde{n}) = \sum_{j=0}^{\infty} A_j^+ \exp\left(\lambda_j^+ \tilde{x}\right) \Phi_{j1}^+(\tilde{n}) \tag{5.67}$$

$$\tilde{x} > 0: \quad \Theta(\tilde{x}, \tilde{n}) = \Psi(\tilde{n}) + (F+1)\tilde{x} - \sum_{j=0}^{\infty} A_j^- \Phi_{j1}^-(\tilde{n}) \exp(\lambda_j^- \tilde{x}) \tag{5.68}$$

with the function $\Psi(\tilde{n})$ given by

$$\Psi(\tilde{n}) = \int_0^{\tilde{n}} \frac{F+1}{\tilde{r}^F a_2(\tilde{n})} \int_0^{\tilde{n}} \tilde{u}(s) \tilde{r}^F \, ds \, d\tilde{n} \; + \; C_2 \; = \; \bar{\Psi}(\tilde{n}) \; + \; C_2 \tag{5.69}$$

where

$$C_2 = \frac{(F+1)^2}{\mathrm{Pe}_L^2} \int_0^1 \tilde{r}^F a_1(\tilde{n}) d\tilde{n} - (F+1) \int_0^1 \tilde{r}^F \tilde{u}(\tilde{n}) \bar{\Psi}(\tilde{n}) d\tilde{n} \tag{5.70}$$

The coefficients A_j are defined by

$$A_j = \frac{\Phi_{j1}(1)}{\lambda_j^2 \|\vec{\Phi}_j\|^2} = -1 \Bigg/ \left[\lambda_j^2 \frac{d}{d\lambda} \left(\frac{\Phi_{j1}'(1)}{\lambda} \right) \right]_{\lambda = \lambda_j} \tag{5.71}$$

It is interesting to focus again on the differences in the solutions for the temperature field for the case of a constant wall temperature and for the case of a constant wall heat flux. For the case of a constant wall temperature, Eqs. (5.40-5.41) show that the temperature in the field changes like an exp-function for positive and negative values of the axial coordinate. Physically, this means that the fluid tries to achieve asymptotically the changed wall temperature after the temperature jump at $\tilde{x} = 0$. For the case of a constant heat flux at the wall, the temperature of the fluid will rise with increasing values of the axial coordinate. This is reflected by the second term on the right hand side of Eq. (5.68). The third term in Eq. (5.68) represents the response of the system to the change in the heat flux for $\tilde{x} = 0$, whereas the first term on the right hand side in Eq. (5.68) represents the fully developed temperature distribution, for large values of the axial coordinate. This behaviour is very similar to the one already discussed in Chap. 3.3. However, there is an important difference, which lies in the fact that the inlet temperature profile (at $\tilde{x} = 0$), for the case with axial heat conduction, is now part of the solution and no longer a boundary condition.

After the temperature distribution in the flow is known, the distribution of the bulk-temperature and the Nusselt number can be calculated from their definitions given by Eqs. (3.64-3.65). This results in

$$\tilde{x} \le 0: \Theta_b(\tilde{x}) = (F+1) \sum_{j=0}^{\infty} A_j^+ \exp\left(\lambda_j^+ \tilde{x}\right) \int_0^1 \tilde{u} \tilde{r}^F \Phi_{j1}^+(\tilde{n}) d\tilde{n} \tag{5.72}$$

$$\tilde{x} > 0 : \Theta_b\left(\tilde{x}\right) = \frac{(F+1)^2}{\mathrm{Pe}_L^2} \int_0^1 \tilde{r}^F a_1\left(\tilde{n}\right) d\tilde{n} + (F+1)\tilde{x} \tag{5.73}$$

$$-(F+1)\sum_{j=0}^{\infty} A_j^- \exp\left(\lambda_j^- \tilde{x}\right) \int_0^1 \tilde{u}\tilde{r}^F \Phi_{j1}^-\left(\tilde{n}\right) d\tilde{n} +$$

$$\tilde{x} > 0 : \mathrm{Nu}_D = -\frac{4}{(F+1)} \left\{ \frac{(F+1)^2}{\mathrm{Pe}_L^2} \int_0^1 \tilde{r}^F a_1\left(\tilde{n}\right) d\tilde{n} - \Psi(1) \right. \tag{5.74}$$

$$\left. -\sum_{j=0}^{\infty} A_j^- \exp\left(\lambda_j^- \tilde{x}\right) \left[(F+1) \int_0^1 \tilde{u}\tilde{r}^F \Phi_{j1}^-\left(\tilde{n}\right) d\tilde{n} - \Phi_{j1}^-(1) \right] \right\}$$

Because of the applied boundary conditions of an adiabatic wall for $\tilde{x} \leq 0$, the Nusselt number has to be zero in this region.

5.2.1 Heat Transfer in Laminar Pipe and Channel Flows for Small Peclet Numbers

For a laminar flow, the functions a_1 and a_2 are equal to one. In addition, the velocity profile is given by a simple parabolic distribution, see Eqs. (3.1-3.2). For this simple case, the function $\Psi(\tilde{n})$, appearing in the temperature distribution, can be predicted analytically (see e.g. Papoutsakis et al. (1980b) for pipe flow). For large values of the axial coordinate ($\tilde{x} \to \infty$), the temperature distribution for $\tilde{x} > 0$, Eq. (5.68), therefore simplifies to

Pipe

$$\Theta(\tilde{x}, \tilde{r}) = 2\tilde{x} + \tilde{r}^2 - \frac{\tilde{r}^4}{4} + \frac{2}{\mathrm{Pe}_L^2} - \frac{7}{24} \tag{5.75}$$

Planar Channel

$$\Theta(\tilde{x}, \tilde{y}) = \tilde{x} + \frac{3}{4}\tilde{y}^2 - \frac{\tilde{y}^4}{12} + \frac{1}{\mathrm{Pe}_L^2} - \frac{1}{7} \tag{5.76}$$

which is the temperature distribution of the fully developed flow.

Laminar Pipe Flow

In order to investigate this case, the flow index F in all the equations has to be set to one. This has been investigated by Papoutsakis et al. (1980b) for the case of a semi-infinite and also for a finite heated zone. Although they derived in their paper equations for the local Nusselt number distribution, only temperature distributions have been shown in their paper. Fig. 5.13 shows the distribution of the bulk-temperature for various values of the Peclet number. This figure elucidates the effect of axial heat conduction. It can be seen that for $Pe_D = 5$, the bulk-temperature is already zero for small distances upstream of $\tilde{x} = 0$. This is different for smaller Peclet numbers. For $Pe_D = 1$, one notices the large effect of the axial heat conduction in the fluid, leading to finite values of Θ_b also far upstream of the entrance into the heating section. The analytically predicted values of the bulk-temperature have been compared with numerical predictions of Hennecke (1968) and very good agreement is found.

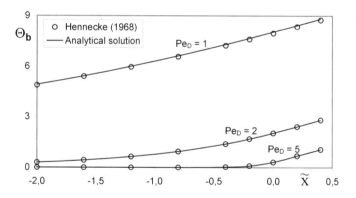

Fig. 5.13: Distribution of the bulk-temperature for different values of the Peclet number (Weigand, Kanzamar and Beer (2001))

Fig. 5.14 shows comparisons between present calculations and predictions by Hsu (1971a), Nguyen (1992), Bilir ((1992) and Hennecke (1968) for two different values of the Peclet number. As it can be seen from Fig. 5.14, the agreement between the present analytical calculations and the Nusselt number from other predictions is very good.

Fig. 5.15 shows the distribution of the local Nusselt number for five different Peclet numbers (data taken from Nguyen (1992)). As it can be seen from Fig. 5.15, the curves for the Nusselt numbers have an inflection point, i.e. Nu_D increases with decreasing Peclet numbers. This phenomenon has been found by various workers, e.g. Hennecke (1968) and Hsu (1971a).

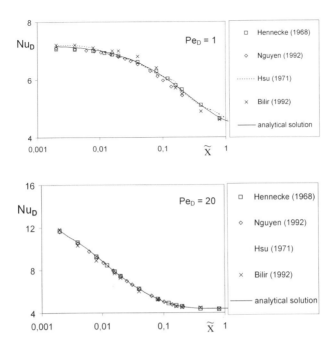

Fig. 5.14: Distribution of the local Nusselt number for two different Peclet numbers

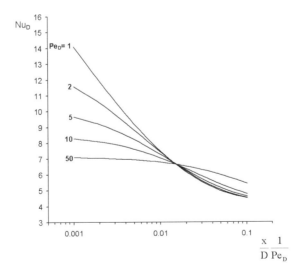

Fig. 5.15: Local Nusselt numbers for pipe flow ($\dot{q}_W = \text{const.}$) and various Pe_D (data taken from Nguyen (1992))

Laminar Channel Flow

In this case, the flow index F has to be set to zero in all equations. Fig. 5.16 shows typical temperature distributions $\Theta_W - \Theta$ for various axial locations. One clearly sees that already for $\tilde{x} = 0$ a temperature profile exists because of the axial heat conduction effects within the fluid. This profile is also present upstream of the start of the heating section. For larger values of the axial coordinate the fully de-veloped temperature distribution, given by Eq. (5.75), is obtained asymptotically.

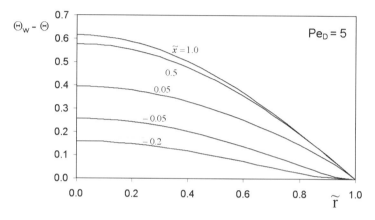

Fig. 5.16: Temperature distribution $\Theta_W - \Theta$ for $Pe_D = 5$ and various axial positions

5.2.2 Heat Transfer in Turbulent Pipe and Channel Flows for Small Peclet Numbers

For a turbulent flow, the functions a_1 and a_2 are depending on the turbulent heat exchange in the flow. For the calculations, assuming a constant wall heat flux boundary condition, the ratio $\varepsilon_{hx} / \varepsilon_{hn}$ can be set to one and the models for the turbulent Prandtl number, explained in paragraph 5.1.2, can be used.

Turbulent Pipe Flow

The general behavior for turbulent pipe flow is similar to the one for laminar flow, even though radial mixing process is enhanced by the turbulent fluctuations. Fig. 5.17 shows distributions of the centerline temperature in the pipe for three differ-ent Peclet numbers and a Prandtl number of 0.001. For a Peclet number as high as 100, it can be seen that no heat is transferred upstream of $\tilde{x} = 0$. Therefore, this problem could be treated as parabolic. For lower values of the Peclet number, ax-ial heat conduction effects start to change the centerline temperature upstream of $\tilde{x} = 0$. Because of the heat transferred upstream, the fluid temperature at the centerline increases to values larger than zero. Fig. 5.18 shows a comparison between analytically predicted Nusselt numbers for the fully developed flow (Nu_{∞}^{-}) and experimental data of Fuchs (1973). It can be seen that the predictions are in good agreement with the experimental data.

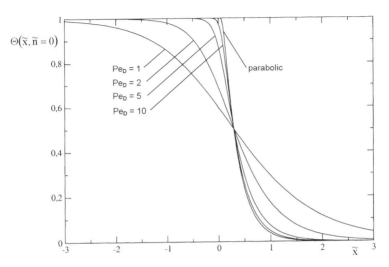

Fig. 5.17: Temperature distribution at the pipe centerline for $Pr = 0.001$

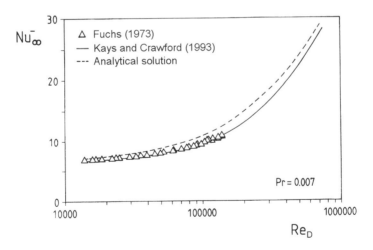

Fig. 5.18: Nusselt number for fully developed flow (Nu_{∞}^{-}) as a function of the Reynolds number (adapted from Weigand, Kanzamar and Beer (1997))

Fig. 5.19 shows the distribution of the local Nusselt number for different Peclet numbers and Pr = 0.001. It can be seen, that increasing values of the Peclet number result in enhanced turbulent mixing and, therefore, in increased values of the Nusselt numbers.

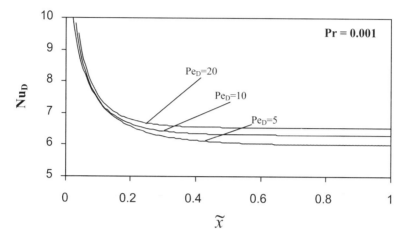

Fig. 5.19: Variation of the local Nusselt number for different Peclet numbers and Pr = 0.001 (Weigand et al. (2001))

Turbulent Channel Flow

For this case, the flow index F has to be set to zero in all equations. Table 5.7 gives the first ten eigenvalues and constants for the turbulent flow in a parallel plate channel for Pr = 0.001 and for two different values of the Reynolds number.

Table 5.7: Eigenvalues and constants for turbulent channel flow with Pr = 0.001.

j	$\text{Re}_D = 5000$		$\text{Re}_D = 10000$	
	λ_j^+	A_j^+	λ_j^+	A_j^+
0	0	0	0	0
1	100.72	2.9632E-4	353.45	4.0451E-7
2	179.41	-2.0154E-3	622.30	-1.9948E-05
3	197.87	4.1893E-3	690.36	2.7807E-04
4	223.14	-3.9981E-3	729.86	-9.6244E-04
5	255.70	2.9457E-3	766.13	1.3601E-03
6	290.88	2.2146E-3	817.95	-1.2109E-03
7	327.40	1.7109E-3	878.26	1.0252E-03
8	36469	-1.3554E-3	943.31	-8.7132E-04
9	402.47	1.0973E-3	1011.5	7.4558E-04
10	440.58	-9.0511E-4	1081.8	-6.4275E-04

j	$-\lambda_j^-$	A_j^-	$-\lambda_j^-$	A_j^-
0	0	0	0	0
1	10.31	-1.7544E-1	10.82	-1.8372 E-1
2	34.50	4.2981E-2	40.14	4.7222 E-2
3	65.28	-1.8317E-2	83.48	-2.0737E-2
4	99.21	9.8721E-3	136.53	1.1391E-2
5	134.85	-6.0627E-3	196.11	-7.0954E-3
6	171.53	4.0512E-3	260.15	4.7916E-3
7	208.88	-2.8746E-3	327.31	-3.4235E-3
8	246.68	2.1344E-3	396.72	2.5503E-3
9	284.80	-1.6420E-3	467.82	-1.9623E-3
10	323.14	1.2997E-3	540.18	1.5496E-3

As it can be seen from the entries in Table 5.7, the first nonzero negative eigenvalue (λ_1^-) stays nearly constant for both Reynolds numbers. On the other side, it can be seen that the first nonzero positive eigenvalue (λ_1^+) increases very strongly with growing values of the Reynolds number. This shows again the decreasing importance of the effect of axial heat conduction with growing values of the Peclet number. This behavior is very similar to the laminar flow case considered before.

5.3 Results for Heating Sections with a Finite Length

In the previous sections, we investigated the axial heat conduction effect for a hydrodynamically fully developed flow for a semi-infinite heating section. For this type of applications, the influence of the axial heat conduction scales very well with the Peclet number, and for flows with Peclet number larger than 100 axial heat conduction effects can be neglected with good accuracy. Furthermore, for the hydrodynamically and thermally fully developed flow, the effect of axial heat conduction can normally be neglected with sufficient accuracy. However, the above given guidelines may not be applicable, if the length of the heating section decreases in size. It is quite obvious that axial heat conduction influences the heat transfer in a very short heated section also for larger values of the Peclet number. In addition, it should be noted that this type of problems can not be calculated correctly using the simplified energy equation (e.g. Eq.(3.14)), which is parabolic in nature. This is obvious because the parabolic nature of the simplified energy equation permits that the end of the heating zone influences the conditions within the heating section. The present section explains how the analytical method previously established can be easily adapted to this sort of problems. The resulting solution is as simple and efficient to compute as the previously discussed solutions for the semi-infinite heating section.

5.3.1 Piecewise Constant Wall Temperature

This case has been recently solved by Lauffer (2003) and Weigand and Lauffer (2004). The geometry under investigation is shown in Fig. 5.20. It can be seen that the wall temperature changes from T_0 to T_W for $0 < x < x_1$. The length of the heating section is x_1.

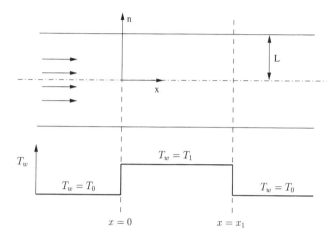

Fig. 5.20: Geometry and boundary condition for a piecewise constant wall temperature

The analysis presented in Chap. 5.1 can be used to solve the present problem. However, the boundary conditions have to be adapted and now are given by

$$n = L : T = T_0, \ x \le 0, \ x \ge x_1 \qquad (5.76)$$
$$T = T_W, 0 < x < x_1$$
$$n = 0 : \partial T / \partial n = 0$$
$$\lim_{x \to -\infty} T = T_o, \qquad \lim_{x \to +\infty} T = T_0$$

The changed boundary conditions result in a changed expression for the function $g(\tilde{x})$, which appears in Eqs. (5.36-5.37). If the integrals in Eqs. (5.36-5.37) are solved, the following temperature distribution in the fluid can be obtained (Lauffer (2003))

$$\tilde{x} \le 0 : \Theta(\tilde{x}, \tilde{n}) = 1 + \sum_{j=0}^{\infty} A_j^+ \Phi_{j1}^+(\tilde{n}) \exp(\lambda_j^+ \tilde{x}) \left[1 - \exp(-\lambda_j^+ \tilde{x}_1) \right] \qquad (5.77)$$

$0 < \tilde{x} < \tilde{x}_1:$

(5.78)

$$\Theta(\tilde{x}, \tilde{n}) = -\sum_{j=0}^{\infty} A_j^- \, \Phi_{j1}^-(\tilde{n}) \exp\left(\lambda_j^- \tilde{x}\right) - \sum_{j=0}^{\infty} A_j^+ \, \Phi_{j1}^+(\tilde{n}) \exp\left(\lambda_j^+ \left(\tilde{x} - \tilde{x}_1\right)\right)$$

$$\tilde{x} \geq \tilde{x}_1 : \Theta(\tilde{x}, \tilde{n}) = 1 - \sum_{j=0}^{\infty} A_j^- \, \Phi_{j1}^-(\tilde{n}) \exp\left(\lambda_j^- \tilde{x}\right)\left[1 - \exp(-\lambda_j^- \tilde{x}_1)\right] \qquad (5.79)$$

where $\tilde{x}_1 = x_1 / (L\,\mathrm{Pe}_L)$. From the above equations, it can be seen that, for $\tilde{x}_1 \to \infty$, the results for the semi-infinite heating section are obtained, (see Eqs. (5.48-5.49)). Furthermore, it is interesting to note that for the case of a finite heated section the temperature distribution within the heating zone is influenced by positive and negative eigenvalues and eigenfunctions (see Eq. (5.78)). The distribution of the Nusselt number can be obtained from the definition of this quantity and the known temperature distribution given by Eqs. (5.77-5.79). This results in

(5.80)

$$\tilde{x} \leq 0 : \mathrm{Nu}_D = \frac{-4 \sum_{j=0}^{\infty} A_j^+ \, \Phi_{j1}'^+(1)\left[\exp(\lambda_j^+ \tilde{x}) - \exp\left(\lambda_j^+ \left(\tilde{x} - \tilde{x}_1\right)\right)\right]}{4^F \sum_{j=0}^{\infty} A_j^+ \int_0^1 \Phi_{j1}^+(\tilde{n}) \tilde{u}\, \tilde{r}^F \, d\tilde{n} \left[\exp(\lambda_j^+ \tilde{x}) - \exp\left(\lambda_j^+ \left(\tilde{x} - \tilde{x}_1\right)\right)\right]}$$

$0 < \tilde{x} < \tilde{x}_1 :$

(5.81)

$$\mathrm{Nu}_D = \frac{-4 \sum_{j=0}^{\infty} A_j^+ \, \Phi_{j1}'^+(1) - 4 \sum_{j=0}^{\infty} A_j^- \, \Phi_{j1}'^-(1) \exp(\lambda_j^- \tilde{x})}{N}$$

$$N = 4^F \sum_{j=0}^{\infty} A_j^+ \int_0^1 \Phi_{j1}^+(\tilde{n}) \tilde{u}\, \tilde{r}^F \, d\tilde{n} \exp\left(\lambda_j^+ \left(\tilde{x} - \tilde{x}_1\right)\right) + \qquad (5.82)$$

$$4^F \sum_{j=0}^{\infty} A_j^- \int_0^1 \Phi_{j1}^-(\tilde{n}) \tilde{u}\, \tilde{r}^F \, d\tilde{n} \exp\left(\lambda_j^- \tilde{x}\right)$$

(5.83)

$$\tilde{x} \geq \tilde{x}_1 : \mathrm{Nu}_D = \frac{-4 \sum_{j=0}^{\infty} A_j^- \, \Phi_{j1}'^-(1)\left[\exp(\lambda_j^- \tilde{x}) - \exp\left(\lambda_j^- \left(\tilde{x} - \tilde{x}_1\right)\right)\right]}{4^F \sum_{j=0}^{\infty} A_j^- \int_0^1 \Phi_{j1}^-(\tilde{n}) \tilde{u}\, \tilde{r}^F \, d\tilde{n} \left[\exp(\lambda_j^- \tilde{x}) - \exp\left(\lambda_j^- \left(\tilde{x} - \tilde{x}_1\right)\right)\right]}$$

The above equations for the distribution of the Nusselt number are identical to the ones given by Eqs. (5.52-5.53) for the case of a semi-infinite heating zone ($\tilde{x}_1 \to \infty$).

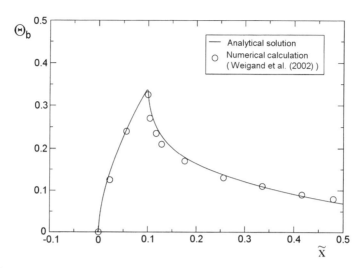

Fig. 5.21: Comparison between numerically and analytically predicted distribution of the bulk-temperature for a turbulent flow in a parallel plate channel with turbulent internal flow ($\text{Re}_D = 40000$, $\text{Pr} = 0.01$, $\tilde{x}_1 = 0.1$)

As mentioned before, the problem of a piecewise heated planar channel has been solved numerically by Weigand et al. (2002). Fig. 5.21 shows a comparison between the analytically and numerically predicted distribution of the bulk-temperature (Lauffer (2003)) for a turbulent flow and a length of the heated zone of $\tilde{x}_1 = 0.1$. It can be seen that the bulk-temperature tries to approach the wall temperature within the heated section. At the end of the heated section, the bulk-temperature decreases then towards the uniform temperature T_0. If the axial coordinate is scaled by the length of the heating zone ($\hat{x} = \tilde{x}/\tilde{x}_1$), the effect of a changing length of the heating zone can be seen nicely. Fig. 5.22 shows the relative error between a parabolic and an elliptic calculation for $\text{Pe}_D = 10$ and a laminar flow through a circular pipe. It can be seen that near the start and the end of the heating section always a large deviation between the elliptic and the parabolic calculation exists. In these areas, the axial heat conduction in the fluid changes the temperature distribution substantially in the pipe and therefore the values of the Nusselt number are changed. If the heating zone decreases in length, axial heat conduction might be important throughout the full heated section. For such a situation, the complete energy equation has to be taken into account. This is clearly visible in Fig. 5.22. For the short heating section of $\tilde{x}_1 = 0.1$, the axial heat conduction in the flow results in a dramatic change of the temperature field, resulting in about 40% relative error in the Nusselt number, even in the middle of the heated section.

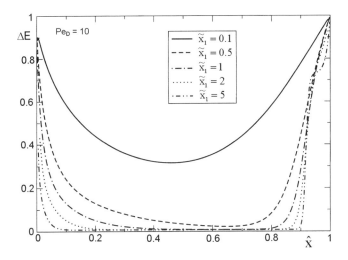

Fig. 5.22: Relative error of the Nusselt number between elliptic and parabolic calculations for a laminar pipe flow ($Pe_D = 10$) with different length of the heating section

5.3.1 Piecewise Constant Wall Heat Flux

The case of a piecewise constant wall heat flux has been discussed by Papoutsakis et al. (1980b) and Weigand et al. (2001). The solution approach is very similar to the one outlined above for the piecewise constant wall temperature. The geometry and the boundary conditions under investigation are shown in Fig. 5.23.

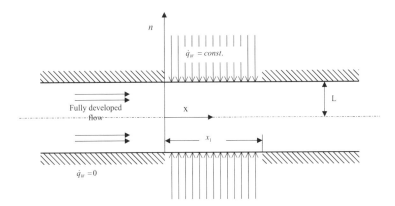

Fig. 5.23: Geometrical configuration and boundary conditions

The boundary conditions given by Eq. (5.57) have to be changed according to

$$n = 0 : \partial T / \partial n = 0,$$
$$n = L : x \le 0, x \ge x_1 : \partial T / \partial n = 0, \quad (5.84)$$
$$0 < x < x_1 : \partial T / \partial n = \dot{q}_w / k$$
$$\lim_{x \to -\infty} T = T_0$$

This results in a changed description of the boundary conditions for E. This means that Eq. (5.58) has to be replaced by

$$\lim_{\tilde{x} \to -\infty} E = 0 \quad , \qquad , \tilde{n} = 0 : E = 0 \quad (5.85)$$

$$\tilde{n} = 1 : E = \begin{cases} 0, & -\infty < \tilde{x} \le 0 \\ \tilde{x}, & 0 < \tilde{x} < \tilde{x}_1 \\ \tilde{x}_1, & \tilde{x}_1 \le \tilde{x} < \infty \end{cases}$$

This changed expression for $E(\tilde{x},1)$ has to be inserted into the integrals in Eqs. (5.63-5.64). These are, however, all the changes we need to do, in order to solve this relatively complicated problem. Evaluating Eqs. (5.63-5.64) results in the following temperature distribution for this case (Weigand et al. (2001))

$$\tilde{x} \le 0 : \Theta(\tilde{x},\tilde{n}) = \sum_{j=0}^{\infty} \frac{\Phi_{j1}^+(1)}{\|\Phi_j^+\|^2} \frac{\exp(\lambda_j^+ \tilde{x})}{\lambda_j^{+2}} \left[1 - \exp(-\lambda_j^+ \tilde{x}_1)\right] \Phi_{j1}^+(\tilde{n}) \qquad (5.86)$$

$$0 < \tilde{x} < \tilde{x}_1 : \Theta(\tilde{x},\tilde{n}) = \Psi(\tilde{n}) + (F+1)\tilde{x} - \sum_{j=0}^{\infty} A_j^- \Phi_{j1}^-(\tilde{n}) \exp(\lambda_j^- \tilde{x}) \qquad (5.87)$$

$$- \sum_{j=0}^{\infty} A_j^+ \Phi_{j1}^+(\tilde{n}) \exp(\lambda_j^+ (\tilde{x} - \tilde{x}_1))$$

$$\tilde{x} \ge \tilde{x}_1 : \Theta(\tilde{x},\tilde{n}) = (F+1)\tilde{x}_1 + \sum_{j=0}^{\infty} A_j^- \Phi_{j1}^-(\tilde{n}) \left[\exp(-\lambda_j^- \tilde{x}_1) - 1\right] \exp(\lambda_j^- \tilde{x}) \qquad (5.88)$$

with the function $\Psi(\tilde{n})$ given by Eq. (5.69). Again the distribution of the bulk-temperature and the Nusselt number can be calculated from the definitions of these quantities and one finally obtains

$$\tilde{x} \le 0 : \Theta_b(\tilde{x}) = (F+1) \sum_{j=0}^{\infty} A_j^+ \left[1 - \exp(-\lambda_j^+ \tilde{x}_1)\right] \exp(\lambda_j^+ \tilde{x}) \int_0^1 \tilde{u} \tilde{r}^F \Phi_{j1}^+(\tilde{n}) d\tilde{n} \qquad (5.89)$$

$$0 < \tilde{x} < \tilde{x}_1 : \Theta_b(\tilde{x}) = \frac{(F+1)^2}{\mathrm{Pe}_L^2} \int_0^1 \tilde{r}^F a_1(\tilde{n}) d\tilde{n} + (F+1)\tilde{x} \qquad (5.90)$$

$$-(F+1)\sum_{j=0}^{\infty} A_j^- \exp\left(\lambda_j^- \tilde{x}\right) \int_0^1 \tilde{u}\tilde{r}^F \Phi_{j1}^-(\tilde{n}) \, d\tilde{n} +$$

$$-(F+1)\sum_{j=0}^{\infty} A_j^+ \exp\left(\lambda_j^+ (\tilde{x}-\tilde{x}_1)\right) \int_0^1 \tilde{u}\tilde{r}^F \Phi_{j1}^+(\tilde{n}) \, d\tilde{n}$$

$\tilde{x} \geq \tilde{x}_1 :$ (5.91)

$$\Theta_b = (F+1)\left\{ \tilde{x}_1 - \sum_{j=0}^{\infty} A_j^- \left[\exp\left(-\lambda_j^- \tilde{x}_1\right) - 1\right] \exp\left(\lambda_j^- \tilde{x}\right) \int_0^1 \tilde{u}\tilde{r}^F \Phi_{j1}^-(\tilde{n}) \, d\tilde{n} \right\}$$

The Nusselt number is zero for $\tilde{x} \leq 0$ and for $\tilde{x} \geq \tilde{x}_1$. For $0 < \tilde{x} < \tilde{x}_1$ the Nusselt number is given by

$$\mathrm{Nu}_D = -\frac{4}{(F+1)}\left\{ \frac{(F+1)^2}{\mathrm{Pe}_L^2} \int_0^1 \tilde{r}^F a_1(\tilde{n}) \, d\tilde{n} - \Psi(1) \right.$$ (5.92)

$$-\sum_{j=0}^{\infty} A_j^- \exp\left(\lambda_j^- \tilde{x}\right)\left[(F+1)\int_0^1 \tilde{u}\tilde{r}^F \Phi_{j1}^-(\tilde{n}) \, d\tilde{n} - \Phi_{j1}^-(1) \right]$$

$$\left. -\sum_{j=0}^{\infty} A_j^+ \exp\left(\lambda_j^+ (\tilde{x}-\tilde{x}_1)\right)\left[(F+1)\int_0^1 \tilde{u}\tilde{r}^F \Phi_{j1}^+(\tilde{n}) \, d\tilde{n} - \Phi_{j1}^+(1) \right] \right\}^{-1}$$

5.4 Application of the Solution Method to Related Problems

The present method can easily be adapted to other related problems. Papoutsakis and Ramkrishna (1981) used the method for example for predicting the heat transfer in a capillary flow emerging from a reservoir. If source terms are present in the energy equation, the method can still provide a complete solution of the problem. The reader is referred to Weigand et al. (1993) for such an application of the method.

The method can also be used to solve related problems for other geometries. Weigand et al. (1997) and Weigand and Wrona (2003) used the method to solve the extended Graetz problem for laminar and turbulent flows inside concentric annuli for constant wall temperature as well as for piecewise constant wall heat flux boundary conditions. The geometry and the boundary conditions are shown in Fig.5.24.

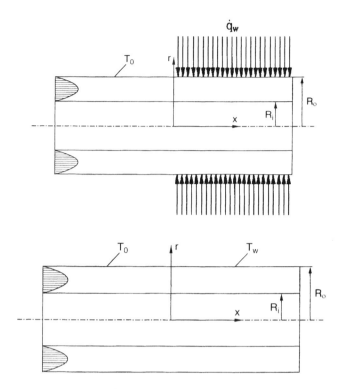

Fig. 5.24: Geometrical configuration and boundary conditions for the extended Graetz problem in a concentric annuli

If another geometry is investigated, the definition of the inner product, given by Eq. (5.14), needs to be adapted. For the heat transfer in concentric annuli it takes the form (see Weigand et al. (1997))

$$\left\langle \vec{\Phi}, \vec{\Upsilon} \right\rangle = (1-\chi) \int_0^1 \left[\frac{a_1(\tilde{r})\big((1-\chi)\tilde{r}+\chi\big)}{Pe_D^2} \Phi_1(\tilde{r}) \Upsilon_1(\tilde{r}) + \frac{\Phi_2(\tilde{r}) \Upsilon_2(\tilde{r})}{4(1-\chi)^2 a_2(\tilde{r})\big((1-\chi)\tilde{r}+\chi\big)} \right] d\tilde{r} \tag{5.93}$$

where the following abbreviations have been used

$$\chi = \frac{R_i}{R_o}, \quad D = 2(R_o - R_i), \quad \tilde{r} = \frac{r - R_i}{R_o - R_i} \tag{5.94}$$

Fig. 5.25 shows, as an example, one result for the heat transfer in a concentric annuli with a constant wall heat flux for laminar flow. The distribution of the Nusselt number is compared with an analytical solution by Hsu (1970), which has been constructed from two independent series solutions for $x < 0$ and $x > 0$. Both, the temperature distribution and the temperature gradient were then matched at $x = 0$ by constructing a pair of orthonormal functions from the eigenfunctions and using the Gram-Schmidt orthonormalization procedure. This method, therefore, was quite cumbersome for finding the solution of the problem. Compared to this solution, the present approach presents the analytical solution in a form, which is as simple as the one for the parabolic problem.

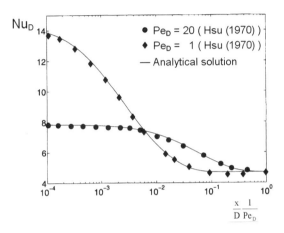

Fig. 5.25: Local Nusselt number for a concentric annuli with constant heat flux at the wall (Weigand and Wrona (2003))

For more details and also for applications for turbulent flows inside concentric annuli the reader is referred to Weigand et al. (1997) and Weigand and Wrona (2003).

Recently, the above described method has also been used for the prediction of the thermally developing laminar flow of a dipolar fluid in a duct by Akyildiz and Bellout (2004). Dipolar fluids are the simplest example of a class of non-Newtonian fluids called multipolar fluids (see Bleustein and Green (1967) for more detailed information on this subject)).

Problems

5-1 Consider the flow and heat transfer in a planar channel with height $2h$. The plates have a constant wall temperature T_0 for $x \leq 0$ and T_W for $x > 0$. The velocity profile of the laminar flow in the planar duct can be simplified to a slug flow profile with $u = \bar{u} = \text{const.}$. The velocity component in the y-direction $v = 0$. All fluid properties can be considered to be constant. Axial heat conduction in the flow can not be neglected, because of small values of the Peclet number. The temperature distribution in the fluid can be calculated from the energy equation

$$u \frac{\partial T}{\partial x} = a \left(\frac{\partial^2 T}{\partial x^2} + \frac{\partial^2 T}{\partial y^2} \right)$$

with the boundary conditions

$$x = 0 : T = T_0$$

$$y = 0 : \frac{\partial T}{\partial y} = 0, \qquad y = h : T = T_W$$

a.) Introduce the following dimensionless quantities

$$\tilde{x} = \frac{x}{h} \frac{a}{\bar{u}h} = \frac{x}{h} \frac{1}{Pe}, \quad \tilde{y} = \frac{y}{h}, \quad \Theta = \frac{T - T_W}{T_0 - T_W}$$

What is the resulting differential equation and the boundary condition in dimensionless form?

b.) Show that the above considered problem reduces in case of negligible axial heat conduction in the fluid ($Pe \rightarrow \infty$) to Problem 3.3.

c.) Use the method of separation of variables to solve the given problem ($\Theta = f(\tilde{x})g(\tilde{y})$). What equations are obtained for the functions f and g?

d.) Predict the complete solution for the temperature field.

e.) Compare the obtained temperature distribution with the one of Problem 3.3 for three different Peclet numbers (Pe = 0.1, 1, 2) for $x/h = 1$. What influence has the axial heat conduction on the temperature profile?

f.) Was it physically correct to prescribe a constant temperature distribution for $x = 0$? Explain your answer.

5-2. Problem 5-1 should now be solved with the method outlined in Chap. 5. The boundary condition for $x = 0 : T = T_0$ is replaced by

$$\lim_{x \to -\infty} T = T_0, \ \lim_{x \to \infty} T = T_W$$

a.) Introduce the dimensionless quantities given in Problem 5.1. What is the resulting dimensionless differential equation and what are the boundary conditions?

b.) Split the problem into two first order partial differential equations by introducing a solution vector $\vec{S}$ consisting of the temperature and the axial energy flow. What system of differential equations is obtained? Show the associated eigenvalue problem and obtain its solution.

c.) Determine the complete solution of the problem.

d.) Compare your solution to the one of Problem 5.1 for $x/h = 1$ for different Peclet numbers. Explain your observations.

5-3. Consideration is given to the heat transfer in a concentric annuli which has constant temperatures at the inner and outer radii for $x \le 0$ and a temperature jump for $x > 0$ at the outer radius. Show from the definition of the inner product (Eq. (5.93)) that

$$\left\langle \vec{\Phi}, \underset{\sim}{L} \tilde{\Upsilon} \right\rangle = \left\langle \underset{\sim}{L} \vec{\Phi}, \tilde{\Upsilon} \right\rangle$$

5-4. Show for the heat transfer in a parallel plate channel that

$$\left\langle \vec{\Phi}_j, \vec{\Phi}_k \right\rangle = 0 \quad \text{for } \lambda_j \ne \lambda_k$$

by considering the eigenvalue problems

$$\underset{\sim}{L} \vec{\Phi}_j = \lambda_j \vec{\Phi}_j \text{ and } \underset{\sim}{L} \vec{\Phi}_k = \lambda_k \vec{\Phi}_k$$

for the case of constant wall temperature.

5-5. Derive from Eq. (5.69-5.70) the fully developed temperature distribution for large axial values for the case of laminar heat transfer in a pipe or a parallel plate channel with constant wall heat flux. Show that Eq. (5.75) is obtained in case of a circular pipe, whereas Eq. (5.76) is obtained for a parallel plate channel.

6 Nonlinear Partial Differential Equations

In the previous chapters, we discussed the solution of linear partial differential equations. Special focus was given to the solution of internal heat transfer problems in duct flows. However, in most technical applications, problems are often described by nonlinear partial differential equations. For a lot of these applications, the equations have to be solved by numerical methods. In contrast to the large amount of literature dealing with the solution of linear partial differential equations, much less literature exists on the solution of nonlinear partial differential equations. One of the major difficulties arising in the solution of nonlinear problems is that we are no longer able to use the powerful superposition method for constructing solutions as for linear problems. Sometimes, the equations under consideration may be linearised by using perturbation methods. An example on how to use this sort of method is shown in Chap. 4 for the solution of eigenvalue problems. In the present chapter, we do not discuss this solution approach. The interested reader is referred to the books of van Dyke (1964), Kevorkian and Cole (1981), Simmonds and Mann (1986), Aziz and Na (1984) and Schneider (1978) for many interesting applications of the perturbation method to fluid dynamics and heat transfer problems. In the following sections, we intend to provide a short overview on some selected solution methods for nonlinear partial differential equations occurring in heat transfer and fluid flow problems. The solution approaches discussed here include, for example, the method of separation of variables, the Kirchhoff transformation, and special solutions of the energy equation. However, the focal point of this chapter is on similarity solutions for the heat conduction and boundary layer equations. Here an overview is given on different methods and on how to determine these solutions. Similarity solutions are not only of importance because they may lead to analytical solutions of the underlying nonlinear partial differential equations, but also, because they play today an important role as excellent benchmark cases for testing computer codes.

6.1 The Method of Separation of Variables

This solution method can also be used for nonlinear partial differential equations. However, since we can not use the superposition principle to fulfill arbitrary boundary conditions, we have to find a solution which already satisfies the partial differential equation and the related boundary conditions. The method is explained for the flow and heat transfer on a rotating disk, which rotates with the constant angular velocity ω in a quiescent fluid. The geometry under investigation is depicted in Fig. 6.1

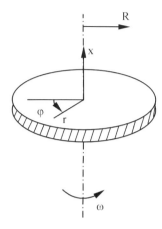

Fig. 6.1: Physical model and coordinate system for the rotating disk

Under the assumptions of angular symmetry for a laminar flow with constant fluid properties and negligible viscous dissipation, the conservation equations in cylindrical coordinates can be written in the following form (see for example Schlichting (1982))

Navier-Stokes Equations

$$v\frac{\partial v}{\partial r} - \frac{w^2}{r} + u\frac{\partial v}{\partial x} = -\frac{1}{\rho}\frac{\partial p}{\partial r} + \frac{\mu}{\rho}\left[\frac{\partial^2 v}{\partial r^2} + \frac{\partial}{\partial r}\left(\frac{v}{r}\right) + \frac{\partial^2 v}{\partial x^2}\right] \tag{6.1}$$

$$v\frac{\partial w}{\partial r} + \frac{wv}{r} + u\frac{\partial w}{\partial x} = \quad \frac{\mu}{\rho}\left[\frac{\partial^2 w}{\partial r^2} + \frac{\partial}{\partial r}\left(\frac{w}{r}\right) + \frac{\partial^2 w}{\partial x^2}\right] \tag{6.2}$$

$$v\frac{\partial u}{\partial r} + u\frac{\partial u}{\partial x} = -\frac{1}{\rho}\frac{\partial p}{\partial x} + \frac{\mu}{\rho}\left[\frac{\partial^2 u}{\partial r^2} + \frac{1}{r}\frac{\partial u}{\partial r} + \frac{\partial^2 u}{\partial x^2}\right] \tag{6.3}$$

Continuity Equation

$$\frac{\partial v}{\partial r} + \frac{v}{r} + \frac{\partial u}{\partial x} = 0 \tag{6.4}$$

Energy Equation

$$v\frac{\partial T}{\partial r}+u\frac{\partial T}{\partial x}=a\left(\frac{\partial^2 T}{\partial r^2}+\frac{1}{r}\frac{\partial T}{\partial r}+\frac{\partial^2 T}{\partial x^2}\right) \tag{6.5}$$

For the rotating disk, the velocity components u, v and w denote the flow in the x, r and φ direction. Furthermore, we assume that the radius R of the rotating disk is very large ($R \to \infty$), so that no boundary conditions have to be satisfied for a finite value of the radius.

The boundary conditions for the above equations are given by

$$x = 0: \quad u = 0, \, v = 0, w = r\omega, T = T_W \tag{6.6}$$

$$x \to \infty : v = 0, w = 0, \quad T = T_\infty$$

No boundary condition is provided for $u(\infty)$, because we only know that this quantity attains a finite value for $x \to \infty$.

As stated above, we consider an incompressible fluid with constant fluid properties. Therefore, the energy equation is decoupled from the momentum equations and the continuity equation. This means that we can first solve the momentum equations and subsequently the energy equation. Let us assume that the solution to Eqs. (6.1-6.4) can be represented by the products

$$v = A(r)F(x) \tag{6.7}$$

$$w = B(r)G(x)$$

$$u = C(r)H(x)$$

From the boundary conditions given by Eq. (6.6), one immediately obtains

$$w = B(r)G(0) = r\omega \tag{6.8}$$

From this equation it follows that $B(r) = B_1\, r$, where B_1 is a constant. If we select this constant to be equal to ω, it follows

$$w = r\,\omega\, G(x), \; G(0) = 1 \tag{6.9}$$

After having gained some knowledge of the tangential velocity component w, let us introduce the expressions given by Eq. (6.9) into the tangential component of the Navier-Stokes equations, Eq. (6.2). This results in

$$v\omega G(x)+vG(x)\omega+urG'(x)\omega = rG''(x)\omega r\mu / \rho \tag{6.10}$$

Introducing also Eq. (6.7) into Eq.(6.10) leads to

$$\omega A(r)F(x)G(x)+\omega A(r)F(x)G(x)+C(r)H(x)r\omega G'(x) = \omega rG''(x)\mu / \rho \tag{6.11}$$

Comparing the left side of this equation to the right side, it is clear that Eq. (6.11) can only have a solution if the function $A(r) = A_1 r$ and $C(r) = C_1$. Selecting the arbitrary constants A_1 and C_1 leads to

$$A(r) = r\omega \tag{6.12}$$
$$C(r) = \sqrt{\omega\mu / \rho}$$

This selection of the constants guarantees that the functions A, B and C have the same dimensions. Summarizing, the velocity field is given by

$$v = r\omega F(x) \tag{6.13}$$
$$w = r\omega G(x)$$
$$u = \sqrt{\omega\mu / \rho}\, H(x)$$

Introducing the above expressions into the momentum equation in the x-direction results in

$$(\omega\mu / \rho) H(x) H'(x) = -\frac{1}{\rho}\frac{\partial p}{\partial x} + \mu / \rho\sqrt{\omega\mu / \rho}\; H''(x) \tag{6.14}$$

From Eq. (6.14) it follows that

$$\frac{\partial p}{\partial x} = p'(x) \quad \rightarrow \quad p = p(x) + I(r) \tag{6.15}$$

Furthermore, one sees from Eq. (6.14) that we can simplify this equation by introducing the following dimensionless coordinate

$$\eta = x\sqrt{\frac{\omega\rho}{\mu}} \tag{6.16}$$

Inserting Eqs. (6.15-6.16) into Eq. (6.14) results in

$$H(\eta)H'(\eta) = -\tilde{p}' + H''(\eta), \quad \tilde{p} = p /(\mu\omega) \tag{6.17}$$

where the prime in Eq. (6.17) indicates the differentiation with respect to η. Introducing Eq. (6.13) and Eqs. (6.15-6.16) into Eq. (6.2) gives

$$\omega^2 F^2 - \omega^2 G^2 + \sqrt{\omega\mu / \rho}\,\omega F' - \mu / \rho\omega F'' = -\frac{1}{\rho}\frac{I'(r)}{r} \tag{6.18}$$

Because on the left hand side of this equation all quantities only depend on the axial coordinate, the right hand side of this equation must be a constant. This leads to the following expression for the function $I(r)$

$$I(r) = C_1 \frac{r^2}{2} + C_2 \tag{6.19}$$

If we examine the pressure distribution for $\eta \to \infty$, the pressure needs to attain a uniform value. This means, that the constant C_1 in Eq. (6.19) must be zero. The constant C_2 can be included into $p(\eta)$ given by Eq. (6.15). Thus, all expressions for the velocity components and the pressure are known. Introducing Eq. (6.13) and Eqs. (6.15-6.16) into the Eqs. (6.1-6.4) and the associated boundary conditions, Eq. (6.6), results in the following set of ordinary differential equations, which has to be solved for obtaining the velocity distribution over the rotating disk

$$F^2 - G^2 + HF' = F'' \tag{6.20}$$

$$2FG + HG' = G'' \tag{6.21}$$

$$HH' = -\tilde{p}' + H'' \tag{6.22}$$

$$2F + H' = 0 \tag{6.23}$$

with the boundary conditions

$$\eta = 0: \quad F = 0, G = 1, H = 0 \tag{6.24}$$
$$\eta \to \infty : F = 0, G = 0$$

The above set of equations has been solved first by von Karman (1921) by using an approximation method. Later Cochran (1934) presented a more accurate solution for the equations.

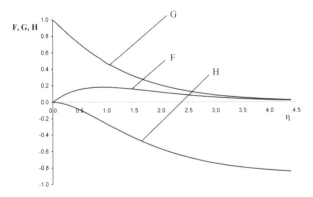

Fig. 6.2: Velocity distribution on a rotating disk (graph compiled from data reported by Cochran (1934))

Figure 6.2 shows the distribution of the functions F, G and H as taken from data reported by Cochran (1934). From Fig. 6.2, it can be seen that the radial and circumferential velocity components decrease rapidly with increasing values of η, while the axial velocity component attains a finite value for large axial distances away from the rotating semi-infinite disk.

Let us now focus on the heat transfer problem for the rotating semi-infinite disk. Introducing the known velocity distribution, Eq. (6.20-6.24), and the dimensionless coordinate, Eq. (6.16), into the energy equation results in

$$\Theta'' = \Pr H \, \Theta' \tag{6.25}$$

where the dimensionless temperature

$$\Theta = \frac{T - T_\infty}{T_W - T_\infty} \tag{6.26}$$

is used. Eq. (6.25) has to be solved together with the following boundary conditions

$$\eta = 0: \quad \Theta = 1 \tag{6.27}$$
$$\eta \to \infty : \Theta = 0$$

The heat transfer problem on a rotating disk has been investigated by Millsaps and Pohlhausen (1952) for Prandtl numbers in the range of 0.5 to 10 and by Sparrow and Gregg (1959) for a large range of different values of the Prandtl number (0.01, 0.1, 1, 10 and 100). Eq. (6.25) has the general solution

$$\Theta = C_1 \int\limits_0^\eta \exp\left(\Pr \int\limits_0^\xi H(\bar{\xi})d\bar{\xi} \right) d\xi + C_2 \tag{6.28}$$

The constants C_1 and C_2 can be obtained by satisfying the boundary conditions according to Eq. (6.27). This results in

$$\Theta = 1 - \int\limits_0^\eta \exp\left(\Pr \int\limits_0^\xi H(\bar{\xi})d\bar{\xi} \right) d\xi \Big/ \int\limits_0^\infty \exp\left(\Pr \int\limits_0^\xi H(\bar{\xi})d\bar{\xi} \right) d\xi \tag{6.29}$$

Sparrow and Gregg (1959) investigated the asymptotic behavior of Eq. (6.29) for small and large values of the Prandtl number. This analysis is shown here in some detail, because it elucidates nicely the behavior of the heat transfer for extreme values of the Prandtl number.

Let us focus first on the case of very small molecular Prandtl numbers. Because the Prandtl number can be interpreted as the ratio of the thickness of the hydrodynamic to the thermal boundary layer (Kays and Crawford (1993)), a very small value of the Prandtl number indicates that the thermal boundary layer is much thicker than the hydrodynamic boundary layer. This means that for very small Prandtl numbers a constant value of the velocity distribution can be used as a first guess in order to solve the convective heat transfer problem. Replacing therefore $H = H(\infty)$ in Eq. (6.25), this equation can be integrated easily. One obtains

$$\Pr \to 0: \Theta = \exp(\Pr H(\infty)\eta), \; H(\infty) = -0.88447 \tag{6.30}$$

The Nusselt number, defined by,

$$\mathrm{Nu} = \frac{h\sqrt{\mu/(\rho\omega)}}{k} = \frac{-k\left(\dfrac{\partial T}{\partial x}\right)_{x=0}\sqrt{\mu/(\rho\omega)}}{k} = -\left(\frac{d\Theta}{d\eta}\right)_{\eta=0} \tag{6.31}$$

where, in Eq. (6.31), the quantity $\sqrt{\mu/(\rho\omega)}$ is used for the length scale in the expression of the Nusselt number. Inserting Eq. (6.30) into Eq. (6.31) results in

$$\Pr \to 0: \mathrm{Nu} = 0.88447\,\Pr \tag{6.32}$$

If the Prandtl number attains very large values, the thermal boundary layer is much thinner than the hydrodynamic boundary layer. This means that, in this case, the velocity distribution can be approximated by a Taylor series expansion of the velocity field around $\eta = 0$. This leads to

$$H = H(0) + H'(0)\eta + \frac{1}{2}H''(0)\eta^2 + \dots \tag{6.33}$$

Since $H(0) = H'(0) = 0$, only the last term in Eq. (6.33) is retained. Inserting this expression into Eq. (6.29) results after integration in

$$\Pr \to \infty : \Theta = 1 - \int_0^\eta \exp\left(\frac{1}{6}\Pr H''(0)\xi^3\right)d\xi \Big/ \int_0^\infty \exp\left(\frac{1}{6}\Pr H''(0)\xi^3\right)d\xi \tag{6.34}$$

For the Nusselt number, one obtains from Eq. (6.34) and Eq. (6.31) (Sparrow and Gregg (1959))

$$\Pr \to \infty : \mathrm{Nu} = \left(-\Pr H''(0)/6\right)^{1/3}/\Gamma(4/3) = 0.62048\,\Pr^{1/3} \tag{6.35}$$

The distribution of the Nusselt number, as a function of the Prandtl number, is shown in Fig. 6.3. The graph is based on calculated values of the Nusselt number from Sparrow and Gregg (1959). In this figure, the asymptotic distributions for small and large values of the Prandtl number are also plotted for comparison. It can be seen that there is a relatively good agreement between the asymptotic values and the numerically calculated Nusselt numbers for extreme values of the Prandtl number.

Before concluding the present example, it is noteworthy stressing once again, where the difference in applying the method of separation of variables to linear and nonlinear partial differential equation lies. The striking difference comes from the fact that the superposition approach can no longer be successfully used. This means that in contrast to Chaps. 2-5, boundary conditions can no longer be satisfied by adding up an infinite set of eigenfunctions.

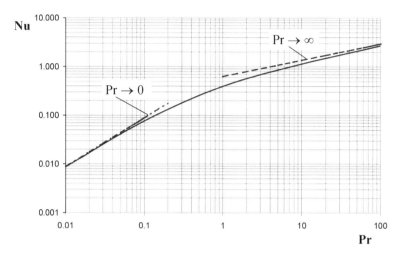

Fig. 6.3: Nusselt numbers for a heated rotating disk for different Prandtl numbers (based on data from Sparrow and Gregg (1959))

6.2 Transformations Resulting in Linear Partial Differential Equations

For some nonlinear partial differential equations, it is possible to reduce them to linear partial differential equations by using special transformations. Some of these transformations are discussed in Ames (1965a), Schneider (1978), Schlichting (1982), Loitsianki (1967), Carslaw and Jaeger (1992) and in Özisik (1968). For elucidating this type of solution approach, we discuss only one case of such a transformation. Consideration is given to the steady-state heat conduction within a two-dimensional region. The temperature field is described by the following heat conduction equation

$$\frac{\partial}{\partial x}\left(k(T)\frac{\partial T}{\partial x} \right) + \frac{\partial}{\partial y}\left(k(T)\frac{\partial T}{\partial y} \right) = 0 \qquad (6.36)$$

The heat conductivity in Eq. (6.36) is assumed to depend on temperature, but not on the coordinates x, y. This is a good assumption, which is sufficiently accurate for a large number of engineering problems. Since the heat conductivity is a function of temperature, the above equation is nonlinear. However, if we define a new dependent variable in the form,

$$\Omega = \int_0^T \frac{k(\tilde{T})}{k_0} d\tilde{T} \qquad (6.37)$$

where k_0 is a reference value of the thermal conductivity, Eq. (6.36) can be transformed into the linear Laplace equation

$$\frac{\partial^2 \Omega}{\partial x^2} + \frac{\partial^2 \Omega}{\partial y^2} = 0 \qquad (6.38)$$

The transformation given by Eq. (6.37) is known in literature as the Kirchhoff
transformation. The only problem arising with this transformation is that also the
boundary conditions have to be in a suitable form, so that after introducing Eq.
(6.37) also the boundary conditions are linear. If we assume constant temperature
boundary conditions, the transformation given by Eq. (6.37) works perfectly and
the resulting equations are linear. The Kirchhoff transformation can also be applied to a much broader class of problems than the one shown above (see for example Carslaw and Jaeger (1992), Özisik (1968)).

Let us consider one special example characterized by the following boundary
conditions for Eq. (6.36)

$$T(0,y) = T_0, \qquad T(l,y) = T_0 \qquad (6.39)$$
$$T(x,0) = T_0, \qquad T(x,l) = T_1$$

where T_0 and T_1 are constant temperatures. Let us further assume that the heat
conductivity within the material is given by the following equation

$$k(T) = k_0 (1 + C(T - T_0)) \qquad (6.40)$$

where k_0 and C are known constants. Introducing the dimensionless quantities

$$\Theta = \frac{T - T_0}{T_1 - T_0}, \; \tilde{x} = \frac{x}{l}, \; \tilde{y} = \frac{y}{l} \qquad (6.41)$$

into Eq. (6.36) and Eq. (6.39) results in

$$\frac{\partial}{\partial \tilde{x}} \left(\frac{k(T)}{k_0} \frac{\partial \Theta}{\partial \tilde{x}} \right) + \frac{\partial}{\partial \tilde{y}} \left(\frac{k(T)}{k_0} \frac{\partial \Theta}{\partial \tilde{y}} \right) = 0 \qquad (6.42)$$

with the boundary conditions

$$\Theta(0,\tilde{y}) = 0, \qquad \Theta(1,\tilde{y}) = 0 \qquad (6.43)$$
$$\Theta(\tilde{x},0) = 0, \qquad \Theta(\tilde{x},1) = 1$$

If we now apply the Kirchhoff transformation

$$\Omega = \int_0^\Theta \frac{k(\tilde{T})}{k_0} d\tilde{T} = \int_0^\Theta (1 + C(T_1 - T_0)\tilde{\Theta}) d\tilde{\Theta} = \Theta + \frac{C(T_1 - T_0)}{2}\Theta^2 \qquad (6.44)$$

to the Eqs. (6.42)-(6.43), we obtain

$$\frac{\partial^2 \Omega}{\partial \tilde{x}^2} + \frac{\partial^2 \Omega}{\partial \tilde{y}^2} = 0 \qquad (6.45)$$

and the corresponding boundary conditions

$$\Omega(0, \tilde{y}) = 0, \qquad \Omega(1, \tilde{y}) = 0 \qquad (6.46)$$

$$\Omega(\tilde{x}, 0) = 0, \qquad \Omega(\tilde{x}, 1) = 1 + \frac{C(T_1 - T_0)}{2}$$

The above given problem for Ω can easily be solved by the method of separation of variables explained in Chap. 2. After obtaining the solution for the function Ω, the temperature distribution for the nonlinear heat conduction problem can be calculated from Eq. (6.44) and is given by

$$\Theta = \frac{1}{C(T_1 - T_0)}\left(\sqrt{2C(T_1 - T_0)\Omega + 1} - 1\right) \qquad (6.47)$$

6.3 Functional Relations Between Dependent Variables

Sometimes it is possible to establish functional relations between different dependent variables. This can be very useful. In case of the flow and heat transfer in a boundary layer it may be used to relate the heat transfer coefficient to the friction factor.

6.3.1 Incompressible Flow over a Heated Flat Plate

In order to elucidate this method, let us investigate the simple problem of a flow over a flat plate. It is assumed that the flow is incompressible and that the fluid properties are constant. The plate surface is kept at a uniform temperature T_W, whereas the fluid far away from the wall moves with the constant velocity u_∞ and has the constant temperature T_∞. The problem under consideration is depicted in Fig. 6.4

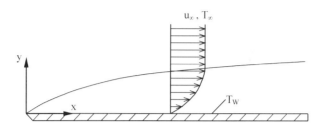

Fig. 6.4: Flow and heat transfer over a flat plate

The problem can be described by the boundary layer equations (Kays and Crawford (1993)) for a laminar flow with constant fluid properties

$$u\frac{\partial u}{\partial x} + v\frac{\partial u}{\partial y} = \frac{\mu}{\rho}\frac{\partial^2 u}{\partial y^2} \tag{6.48}$$

$$\frac{\partial u}{\partial x} + \frac{\partial v}{\partial y} = 0 \tag{6.49}$$

$$u\frac{\partial T}{\partial x} + v\frac{\partial T}{\partial y} = a\frac{\partial^2 T}{\partial y^2} \tag{6.50}$$

with the boundary conditions

$$y = 0: \quad u = 0, \ v = 0, \ T = T_W \tag{6.51}$$
$$y \to \infty : u = u_\infty, \qquad T = T_\infty$$

By comparing Eq. (6.48) to Eq. (6.50), it can be seen that both equations have a very similar structure. Introducing the following dimensionless quantities

$$\tilde{u} = \frac{u}{u_\infty}, \tilde{v} = \frac{v}{u_\infty}, \tilde{x} = \frac{x}{\delta}, \tilde{y} = \frac{y}{\delta}, \Theta = \frac{T - T_W}{T_\infty - T_W} \tag{6.52}$$

into Eqs. (6.48-6.51), where δ denotes a length scale, one obtains

$$\tilde{u}\frac{\partial \tilde{u}}{\partial \tilde{x}} + \tilde{v}\frac{\partial \tilde{u}}{\partial \tilde{y}} = \frac{\mu}{\rho u_\infty \delta}\frac{\partial^2 \tilde{u}}{\partial \tilde{y}^2} \tag{6.53}$$

$$\frac{\partial \tilde{u}}{\partial \tilde{x}} + \frac{\partial \tilde{v}}{\partial \tilde{y}} = 0 \tag{6.54}$$

$$\tilde{u}\frac{\partial\Theta}{\partial\tilde{x}} + \tilde{v}\frac{\partial\Theta}{\partial\tilde{y}} = \frac{a}{u_\infty\delta}\frac{\partial^2\Theta}{\partial\tilde{y}^2} \tag{6.55}$$

with the boundary conditions

$$\tilde{y} = 0: \ \tilde{u} = 0, \ \tilde{v} = 0, \ \Theta = 0 \tag{6.56}$$
$$\tilde{y} \to \infty: \tilde{u} = 1, \qquad \Theta = 1$$

If we finally assume that the Prandtl number of the fluid is 1, which means that $a = \mu/\rho$ (which is a good assumption for all gases), one can see that Eq. (6.53) is identical to Eq. (6.55). In addition, the boundary conditions for $\tilde{u}$ and Θ are the same. This means that both equations must have the same solution. Thus

$$\Theta = \tilde{u}, \quad \frac{T - T_W}{T_\infty - T_W} = \frac{u}{u_\infty} \tag{6.57}$$

Because the Prandtl number is equal to 1, the thickness of the hydrodynamic and the thermal boundary layer is the same in size. Equation (6.57) states the simple fact that the velocity profile in the boundary layer has in such a case the same shape as the temperature profile. Equation (6.57) is a very useful equation, because it can be used to establish a relation between the friction factor and the heat transfer coefficient at the surface.

In order to show this, we will first predict the heat flux at the surface

$$\dot{q}_W = -k\left(\frac{\partial T}{\partial y}\right)_W = -k\left(\frac{\partial T}{\partial u}\right)\left(\frac{\partial u}{\partial y}\right)_W = -\frac{k(T_\infty - T_w)}{u_\infty}\left(\frac{\partial u}{\partial y}\right)_W \tag{6.58}$$

Introducing the friction factor c_f defined by

$$c_f = \frac{2\tau_W}{\rho u_\infty^2} = \mu\left(\frac{\partial u}{\partial y}\right)_W \frac{2}{\rho u_\infty^2} \tag{6.59}$$

results in

$$\dot{q}_W = \frac{k(T_W - T_\infty)}{\mu}\frac{c_f}{2}\rho u_\infty \tag{6.60}$$

If we finally introduce the definition of the Nusselt number, based on the plate length L

$$\mathrm{Nu}_L = \frac{hL}{k} = \frac{\dot{q}_W L}{k(T_\infty - T_W)} \tag{6.61}$$

into the above equation, one obtains

$$\frac{\mathrm{Nu}_L}{\mathrm{Re}_L \, \mathrm{Pr}} = \mathrm{St} = \frac{c_f}{2} \qquad (6.62)$$

Equation (6.62) states the interesting fact that the knowledge of the friction factor is sufficient for predicting the heat transfer for the above case. This relation is known as the Reynolds analogy in convective heat transfer. It is traditionally derived for a turbulent flow with $\mathrm{Pr} = 1$ and $\mathrm{Pr}_t = 1$ (see for example Kays and Crawford (1993)). The equation can be used in order to obtain an approximate value of the heat transfer coefficient by simply using the known friction factor.

6.3.2 Compressible Flow over a Flat, Heated Plate

Having established a direct relation between the friction factor and the heat transfer coefficient, one could ask whether similar expressions between the temperature field and the velocity field exist also for compressible flows. This is indeed the case and the following analysis should serve as another example for establishing functional dependences between dependent variables. Let us consider the case of a compressible flow over a flat plate. The equations describing the laminar flow and heat transfer in this situation are the compressible boundary layer equations (Kays and Crawford (1993), Schlichting (1982), Jischa (1982)). For the case of no axial pressure gradient ($dp/dx = 0$) in the flow, these equations are given by

$$\frac{\partial}{\partial x}(\rho u) + \frac{\partial}{\partial y}(\rho v) = 0 \qquad (6.63)$$

$$\rho \left(u \frac{\partial u}{\partial x} + v \frac{\partial u}{\partial y} \right) = \frac{\partial}{\partial y}\left(\mu \frac{\partial u}{\partial y} \right) \qquad (6.64)$$

$$\rho c_p \left(u \frac{\partial T}{\partial x} + v \frac{\partial T}{\partial y} \right) = \frac{\partial}{\partial y}\left(k \frac{\partial T}{\partial y} \right) + \mu \left(\frac{\partial u}{\partial y} \right)^2 \qquad (6.65)$$

In addition to the conservation equations, we have to prescribe for a compressible flow also a relation between p, ρ and T. This is accomplished by the thermal state equation. If we assume that the fluid can be treated as an ideal gas, one has

$$p = \rho R T \qquad (6.66)$$

The above equations have to be solved with the boundary conditions

$$y = 0: \quad u = 0, \ v = 0, \ T = T_w \qquad (6.67)$$

$$y \rightarrow \infty : u = u_\infty, \qquad T = T_\infty$$

For a compressible flow, the Eqs. (6.63-6.66) are strongly coupled by the varying density. As in the case of the incompressible flow, we are looking for a functional relationship of the form

$$T = f(u) \tag{6.68}$$

Inserting Eq. (6.68) into the energy equation, Eq. (6.65), results in

$$\rho c_p \frac{dT}{du}\left(u\frac{\partial u}{\partial x} + v\frac{\partial v}{\partial y} \right) = \frac{\partial}{\partial y}\left(k\frac{\partial u}{\partial y}\frac{dT}{du} \right) + \mu\left(\frac{\partial u}{\partial y} \right)^2 \tag{6.69}$$

This equation can be further simplified using Eq. (6.64) and one obtains

$$c_p \frac{dT}{du}\frac{\partial}{\partial y}\left(\mu\frac{\partial u}{\partial y} \right) = \frac{\partial}{\partial y}\left(k\frac{\partial u}{\partial y}\frac{dT}{du} \right) + \mu\left(\frac{\partial u}{\partial y} \right)^2 \tag{6.70}$$

Performing all differentiations on the right hand side of Eq. (6.70) leads to

$$c_p \frac{dT}{du}\frac{\partial}{\partial y}\left(\mu\frac{\partial u}{\partial y} \right) = \frac{dT}{du}\frac{\partial}{\partial y}\left(k\frac{\partial u}{\partial y} \right) + k\frac{d^2 T}{du^2}\left(\frac{\partial u}{\partial y} \right)^2 + \mu\left(\frac{\partial u}{\partial y} \right)^2 \tag{6.71}$$

Rearranging gives finally

$$\frac{dT}{du}\left(c_p \frac{\partial}{\partial y}\left(\mu\frac{\partial u}{\partial y} \right) - \frac{\partial}{\partial y}\left(k\frac{\partial u}{\partial y} \right) \right) = \left(\frac{\partial u}{\partial y} \right)^2\left(k\frac{d^2 T}{du^2} + \mu \right) \tag{6.72}$$

Let us now assume that c_p is constant. Then we introduce the Prandtl number ($\text{Pr} = \mu c_p / k$) into Eq. (6.72). This leads to

$$\frac{dT}{du}\left(\frac{\partial}{\partial y}\left(k\,\text{Pr}\frac{\partial u}{\partial y} \right) - \frac{\partial}{\partial y}\left(k\frac{\partial u}{\partial y} \right) \right) = \left(\frac{\partial u}{\partial y} \right)^2\left(k\frac{d^2 T}{du^2} + \mu \right) \tag{6.73}$$

For Pr = 1, Eq. (6.73) simplifies drastically and one obtains

$$k\frac{d^2 T}{du^2} + \mu = 0 \tag{6.74}$$

Equation (6.74) is an ordinary differential equation, which can be integrated easily. Adapting the solution to the boundary conditions given by Eq. (6.67) leads to

$$\frac{T - T_W}{T_\infty - T_W} = \frac{u}{u_\infty} + \frac{u_\infty^2}{2c_p(T_\infty - T_W)}\frac{u}{u_\infty}\left(1 - \frac{u}{u_\infty} \right) \tag{6.75}$$

Equation (6.75) shows nicely that the linear dependence between temperature and velocity, see Eq. (6.57), is now replaced by a nonlinear relationship. For small velocities, the quadratic term on the right hand side of Eq. (6.75) can be neglected and one obtains again Eq. (6.57). However, if we rewrite Eq. (6.75) using the stagnation temperatures instead of static temperatures, which means

$$T_0 = T + \frac{1}{2}\frac{u^2}{c_p}, \quad T_{0\infty} = T_\infty + \frac{1}{2}\frac{u_\infty^2}{c_p}, \quad T_{0W} = T_W \tag{6.76}$$

one obtains again

$$\frac{T_0 - T_{0W}}{T_{0\infty} - T_{0W}} = \frac{u}{u_\infty} \tag{6.77}$$

This is an extension of the similarity between the velocity distribution and the temperature distribution given by Eq. (6.57). For a more detailed discussion of this case, the reader is referred to Schlichting (1982) and Loitsianki (1967).

6.4 Similarity Solutions

In the following section, we are interested in "similar solutions" of nonlinear partial differential equations. Mathematically speaking, similarity solutions of partial differential equations appear when the number of independent variables can be reduced by a transformation of these variables. If we focus on the important case of a partial differential equation with two independent variables, this means that an ordinary differential equation is obtained after the transformation.

Similarity solutions play an important role in boundary layer theory and also in heat conduction. For example, in boundary layer theory, similarity solutions can be obtained for some cases if the physical problem under consideration is lacking of a characteristic length or a characteristic time scale (e.g. the boundary layer on a semi-infinite plate with zero pressure gradient). In such cases, the nonlinear partial differential equations for the boundary layer are reduced to nonlinear ordinary differential equations. There are several books dealing with similarity solutions and the different methods to obtain them. The reader is referred for example to Hansen (1964), Ames (1965a), Schneider (1978), Dewey and Gross (1967) and Özisik (1968) for detailed discussions on similarity solutions. Basically, there are three main classes of methods for obtaining similarity solutions (Hansen (1964)). These three methods are explained in the following section for a simple heat conduction problem. Subsequently, we analyse similarity solutions for different boundary layer problems.

6.4.1 Similarity Solutions for a Transient Heat Conduction Problem

Consider the one-dimensional transient heat conduction in a semi-infinite solid. The body has a uniform initial temperature T_0. For $t \geq 0$, the wall temperature is set to the constant temperature T_W. The properties of the solid are assumed to be constant. The problem is described by the heat conduction equation for the solid

$$\rho c \frac{\partial T}{\partial t} = \frac{\partial}{\partial x} \left(k \frac{\partial T}{\partial x} \right)$$ (6.78)

with the boundary conditions

$$t = 0 : \; T = T_0$$ (6.79)
$$x = 0 : \; T = T_W , \quad x \to \infty : T = T_0$$

The problem under consideration is depicted in Fig. 6.5. Of course, Eq. (6.78) represents a linear partial differential equation and can be solved also with other methods. However, the simplicity of this equation makes it a good candidate to explain the different methods for obtaining similarity solutions, before considering more complicated, nonlinear problems.

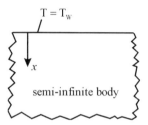

Fig. 6.5: Transient heat conduction in a semi-infinite solid

Methods Based on Dimensional Analysis

The dimensional analysis is a powerful method, which can be used for all physical applications. The method is based on the fact, that all physical quantities have units, which can be split into "basic units". These basic units could be, for example, the international SI-system with the units: Length L (m), mass M (kg), time τ (s), temperature ϑ (K), species amount N (mol), current strength I (A) and light strength S (cd). The density for example has the unit mass M divided by length cubed L^3, which can be denoted by $[ML^{-3}]$.

In general, one can say that for any problem there is always the task to express a physical quantity p_1 as a function of other physical quantities $p_2, p_3, ..., p_n$, where n is the number of physical variables needed to describe the problem under consideration (Görtler (1975), Spurk (1992), Simon (2004)). This means that a relation

$$f(p_1, p_2, ..., p_n) = 0$$ (6.80)

can be established. Each of the physical variables p_j has units, which can be constructed from the basic units stated above. This means that the dimension of the variable p_j, denoted by $[p_j]$, is given by

$$[p_j] = \prod_{i=1}^{m} (X_i)^{a_{ij}}$$ (6.81)

where X_i denotes one of the basic units powered by the exponent a_{ij}. For the above example of the density, one has

$$[\rho] = M^1 L^{-3}, \ a_1 = 1, \ a_2 = -3$$ (6.82)

Since every physical process must be independent of the arbitrary units used, the relation given by Eq. (6.80) can be reduced to a relationship between dimensionless quantities Π_j, so that.

$$f(\Pi_1, \Pi_2, ..., \Pi_d) = 0$$ (6.83)

where d is the number of dimensionless products. These dimensionless quantities are products of the physical quantities, which means

$$\Pi = \prod_{j=1}^{n} (p_j)^{b_j}$$ (6.84)

with the dimension

$$[\Pi] = 1 = \prod_{j=1}^{n} [p_j]^{b_j}$$ (6.85)

Inserting Eq. (6.81) into Eq. (6.85) leads to

$$[\Pi] = 1 = \prod_{j=1}^{n} \prod_{i=1}^{m} (X_i)^{a_{ij} b_j}$$ (6.86)

In Eq. (6.86) the coefficients a_{ij} are known and the exponents b_j have to be determined, so that the equation is satisfied. This is always possible, if the system of equations

$$\sum_{j=1}^{n} a_{ij} b_j = 0$$ (6.87)

has nontrivial solutions for the n unknown coefficients b_j. The coefficient matrix a_{ij} is also known as the dimension matrix.

The equivalence between Eq. (6.80) and Eq. (6.83) is known in literature as the "Buckingham theorem" or as the "Pi theorem" (Görtler (1978), Spurk (1992), Simon (2004)). This theorem gives also a relation between the number of physical variables involved, n, and the number of independent dimensionless groups, d, according to

$$d = n - r \tag{6.88}$$

where r is the rank of the dimension matrix.

Let us now focus again on the problem of finding similarity solutions for the Eqs. (6.78-6.79). The physical quantities involved in the problem are: $\rho, c, k, T_0, T_W, T, t, x$. This means that we can construct the following dimension matrix

Table 6.1: Dimension matrix for the transient heat conduction problem

	ρ	c	k	$T - T_0$	$T_0 - T_W$	t	x
L	-3	2	1	0	0	0	1
M	1	0	1	0	0	0	0
τ	0	-2	-3	0	0	1	0
ϑ	0	-1	-1	1	1	0	0

where the units for the heat conductivity $k\ \left[Wm^{-1}K^{-1} \right]$ and for the specific heat c $\left[J\,kg^{-1}K^{-1} \right]$ have been expressed in basic units. Note that, in the dimension matrix, we have used only temperature differences. The rank of the above dimension matrix $r = 4$, which implies that the number of dimensionless products is $d = n - r = 7 - 4 = 3$. However, upon inspection of Eq. (6.78), one notices that the quantities ρ, c, k do not appear independently in this equation. They can be lumped together to build the heat diffusivity $a = k /(\rho c)$. By doing so, the dimension matrix simplifies to

Table 6.2: Simplified dimension matrix for the transient heat conduction problem

	a	$T - T_0$	$T_0 - T_W$	t	x
L	2	0	0	0	1
M	0	0	0	0	0
τ	-1	0	0	1	0
ϑ	0	1	1	0	0

In this matrix, no entries appear for the mass M, so that the rank of this matrix is only 3. This means that we can construct 2 dimensionless quantities. By doing so, one obtains

Table 6.3: Dimension matrix with dimensional groups

	$x/\sqrt{at}$	$(T-T_0)/(T_W-T_0)$	T_0-T_W	t	x
L	0	0	0	0	1
M	0	0	0	0	0
τ	0	0	0	1	0
ϑ	0	0	1	0	0

This means that the solution of the current problem must have the functional form

$$\frac{T-T_0}{T_W-T_0} = f\left(\frac{x}{\sqrt{at}}\right) \tag{6.89}$$

Introducing the new coordinate

$$\eta = \frac{x}{\sqrt{at}} \tag{6.90}$$

into Eqs. (6.78-6.79), results in the ordinary differential equations

$$\Theta'' + \frac{1}{2}\eta\Theta' = 0 \tag{6.91}$$

where the prime denotes differentiation with respect to the coordinate η. The boundary conditions for this equation are

$$\eta = 0: \quad \Theta = 1 \tag{6.92}$$
$$\eta \to \infty: \Theta = 0$$

Equation (6.92) shows nicely that the two boundary conditions for $t = 0$ and for $x \to \infty$ given by Eq. (6.79) have been replaced by one boundary condition for $\eta \to \infty$. Eq. (6.91) together with the boundary conditions given by Eq. (6.92) can be integrated and one finally obtains

$$\Theta = 1 - \frac{1}{\sqrt{\pi}}\int_0^{\eta}\exp\left(-\frac{\bar{\eta}^2}{4}\right)d\bar{\eta} \tag{6.93}$$

Group-Theory Methods

This method for obtaining similarity solutions has been suggested by Birkhoff (1948) and has been investigated extensively by Morgan (1951), Müller and Matschat (1962), Aimes (1965a) and Aimes (1965b). The method implies that the search for similarity solutions for a system of partial differential equations is equivalent to the determination of the invariant solutions of these equations under a particular parameter group of transformation. A set of similarity variables, which are invariants of the group, is obtained by solving a set of simultaneous algebraic equations.

Suppose Γ is a set of N partial differential equations given by

$$\Gamma : \psi_j(x_i, y_j) = 0, \quad i = 1, 2, 3, ..., M \tag{6.94}$$
$$j = 1, 2, 3, ..., N$$

where x_i, $i = 1, 2, 3, ..., M$ are the independent and y_j, $j = 1, 2, 3, ..., N$ are the dependent variables. Now we are considering a one-parameter group G_1 with the nonzero real parameter A, whose elements are given by

$$\overline{x}_i = A^{\gamma_i} x_i \tag{6.95}$$
$$\overline{y}_j = A^{\kappa_j} y_j$$

In Eq. (6.95) A is a nonzero real number, called the parameter of the transformation, and the exponents γ_i, κ_j have to be determined from the requirement that the system of partial differential equations, given by Eq. (6.94), is absolutely invariant under the transformation given by Eq. (6.95). This means that

$$\psi_j(x_i, y_j) = \psi_j(\overline{x}_i, \overline{y}_j) \tag{6.96}$$

After applying the transformation to the partial differential equations, one obtains a set of simultaneous algebraic equations for determining the unknown coefficients γ_i and κ_j. More detailed information on this method can be found in the literature mentioned above.

Let us illustrate the method by obtaining similarity solutions for the Eqs. (6.78-6.79). We assume a linear transformation for the dependent and independent variables in Eq. (6.76) of the form

$$\overline{x} = A^{\gamma_1} x, \ \overline{t} = A^{\gamma_2} t \tag{6.97}$$
$$\overline{T} = A^{\kappa_1} T$$

As noted before $\gamma_1, \gamma_2, \kappa_1$ are constants and the quantity A is called the parameter of the transformation. Solving the above given equation for A gives

$$A = \left(\frac{\overline{x}}{x}\right)^{1/\gamma_1} = \left(\frac{\overline{t}}{t}\right)^{1/\gamma_2} = \left(\frac{\overline{T}}{T}\right)^{1/\kappa_1} \tag{6.98}$$

Introducing the quantities given by Eq. (6.97) into Eq. (6.78) results in

$$A^{\gamma_2 - \kappa_1} \rho c \frac{\partial \overline{T}}{\partial \overline{t}} = A^{2\gamma_1 - \kappa_1} \frac{\partial}{\partial \overline{x}} \left(k \frac{\partial \overline{T}}{\partial \overline{x}}\right) \tag{6.99}$$

Comparing Eq. (6.99) to Eq. (6.78) shows clearly that the partial differential equation does not get altered by the transformation if

$$\kappa_1 - \gamma_2 = \kappa_1 - 2\gamma_1 \tag{6.100}$$

This equation leads immediately to the result that $\gamma_2 = 2\gamma_1$, whereas nothing can be said about κ_1. This is obvious, because the differential equation is linear and therefore every exponent for the temperature drops out. Introducing the relationship between γ_1 and γ_2 into Eq. (6.98), it can be seen that the coordinates should be related in the following way in order to guarantee similarity solutions

$$\left(\frac{\bar{x}}{x}\right)^{1/\gamma_1} = \left(\frac{\bar{t}}{t}\right)^{1/2\gamma_1} \Rightarrow \bar{\eta} = \frac{x}{\sqrt{t}} \tag{6.101}$$

This new coordinate can be made dimensionless by using $a = k/(\rho c)$ and one obtains the similarity coordinate given by Eq. (6.90)

$$\eta = \frac{x}{\sqrt{at}} \tag{6.90}$$

Introducing again the dimensionless temperature $\Theta = (T - T_0)/(T_W - T_0)$ leads after performing the change of variables in Eq. (6.78), to the ordinary differential equation (6.91) with boundary conditions given by Eq. (6.92).

Separation of Variables and the Method of the Free Variable

Both methods are strongly related to each other. We have already seen in section 6.1 how the method of separation of variables can be used in order to find a similarity solutions for partial differential equations. Here we want to investigate the method of the free variable (see also Schneider (1978)). For this method, we simply transform the coordinates in Eq. (6.78) according to

$$\tau = t \tag{6.102}$$
$$\eta = x\,g(t)$$

The new variable η contains the unknown function $g(t)$. This variable is the free variable, which gave the method its name. In order to use the method for obtaining similarity solutions for Eqs. (6.78-6.79), we introduce the dimensionless temperature distribution $\Theta = (T - T_0)/(T_W - T_0)$ into these equations. This results in

$$\frac{\partial \Theta}{\partial t} = a\frac{\partial^2 \Theta}{\partial x^2} \tag{6.103}$$

with the boundary conditions

$$t = 0: \ \Theta = 0 \tag{6.104}$$
$$x = 0: \Theta = 1, \quad x \to \infty: \Theta = 0$$

Introducing the new coordinates defined in Eq. (6.102) into the partial differential equation (6.103) results in

$$\frac{\eta}{g(\tau)} g'(\tau) \frac{d\Theta}{d\eta} = a g^2(\tau) \frac{d^2\Theta}{d\eta^2}$$ (6.105)

where the prime denotes differentiation with respect to τ. In Eq. (6.105) we have assumed that Θ depends only on η. This assumption is suggested by the form of the boundary conditions given by Eq. (6.104). Eq. (6.105) reduces to an ordinary differential equation, if all dependence on the time drops out of this equation. This is satisfied, if

$$\frac{g'(\tau)}{g^3(\tau)} = \text{const.}$$ (6.106)

From this equation we can calculate $g(\tau)$

$$g(\tau) = 1/\sqrt{C_1 + a\tau}$$ (6.107)

The constant C_1 in Eq. (6.107) can be set to zero, because it only shifts the time origin and we are only interested in one solution of Eq. (6.106). Introducing Eq. (6.107) with $C_1 = 0$ into Eq. (6.105) leads to

$$\Theta'' + \frac{1}{2}\eta\Theta' = 0$$ (6.91)

which is identical to Eq. (6.91).

6.4.2 Similarity Solutions of the Boundary Layer Equations for Laminar Free Convection Flow on a Vertical Flat Plate

After discussing the fundamentals of the different methods for a very simple example, the present section investigates similarity solutions of the boundary layer equations for laminar, incompressible, free convection flow on a vertical flat wall. The reader is referred to Jaluria (1980) and Ede (1967) for detailed discussions on this topic. Similarity solutions for this type of application have been obtained for example by Ostrach (1953) which are in very good agreement with measurements performed by Schmidt and Beckmann (1930).

The geometry as well as the velocity distribution for the free convection on a vertical flat plate is shown in Fig. 6.6.

The free convection flow is driven by the temperature difference between the wall and the free-stream. It is assumed that the Bousinesq approximation can be applied. This means that the density difference can be replaced by a simplified equation of state and that variable property effects can be neglected, except for the density in the momentum equation (see for example Jaluria (1980)).

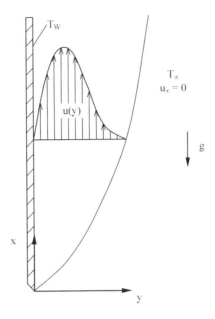

Fig. 6.6: Free convection flow on a vertical flat plate

In the above equation, g is the gravitational acceleration and β denotes the thermal volumetric expansion coefficient, defined by

$$\beta = \frac{1}{\rho}\left(\frac{\partial \rho}{\partial T}\right)_{p}$$
(6.109)

The continuity equation (6.49) can automatically be satisfied by introducing a stream function, defined by

$$u = \frac{\partial \Psi}{\partial y} \ , \ v = -\frac{\partial \Psi}{\partial x}$$
(6.110)

This leads to the following set of partial differential equations

$$\frac{\partial \Psi}{\partial y}\frac{\partial^{2}\Psi}{\partial x \partial y} - \frac{\partial \Psi}{\partial x}\frac{\partial^{2}\Psi}{\partial y^{2}} = g\beta(T_{W} - T_{\infty})\Theta + \frac{\mu}{\rho}\frac{\partial^{3}\Psi}{\partial y^{3}}$$
(6.111)

$$\frac{\partial \Psi}{\partial y}\frac{\partial \Theta}{\partial x} - \frac{\partial \Psi}{\partial x}\frac{\partial \Theta}{\partial y} = a\frac{\partial^{2}\Theta}{\partial y^{2}}$$
(6.112)

with boundary conditions

$$y = 0 \quad : \partial \Psi / \partial y = 0, \ \partial \Psi / \partial x = 0, \ \Theta = 1 \tag{6.113}$$
$$y \to \infty : \partial \Psi / \partial y = 0, \qquad\qquad \Theta = 0$$

and the dimensionless temperature

$$\Theta = \frac{T - T_\infty}{T_W - T_\infty} \tag{6.114}$$

Group-Theory Methods

The group-theory method has been used, for example, by Ames (1965a) and by Cebeci and Bradshaw (1984) for obtaining similarity solutions for the boundary layer equations for forced convection flows. Here we want to investigate the free convection on a vertical plate by using the linear transformation

$$\bar{x} = A^{\gamma_1} x, \qquad \bar{\Psi} = A^{\kappa_1} \Psi \tag{6.115}$$
$$\bar{y} = A^{\gamma_2} y, \qquad \bar{\Theta} = A^{\kappa_2} \Theta$$

for the dependent and independent variables. From Eq. (6.115) one obtains

$$\left(\frac{\bar{x}}{x} \right)^{1/\gamma_1} = \left(\frac{\bar{y}}{y} \right)^{1/\gamma_2} = \left(\frac{\bar{\Psi}}{\Psi} \right)^{1/\kappa_1} = \left(\frac{\bar{\Theta}}{\Theta} \right)^{1/\kappa_2} = A \tag{6.116}$$

Introducing the linear transformation, given by Eq. (6.115), into the Eqs. (6.111-6.112) results in

$$A^{\gamma_1 + 2\gamma_2 - 2\kappa_1} \frac{\partial \bar{\Psi}}{\partial \bar{y}} \frac{\partial^2 \bar{\Psi}}{\partial \bar{x} \partial \bar{y}} - A^{\gamma_1 + 2\gamma_2 - 2\kappa_1} \frac{\partial \bar{\Psi}}{\partial \bar{x}} \frac{\partial^2 \bar{\Psi}}{\partial \bar{y}^2} = \tag{6.117}$$

$$g \beta (T_W - T_\infty) A^{-\kappa_2} \bar{\Theta} + A^{3\gamma_2 - \kappa_1} \frac{\mu}{\rho} \frac{\partial^3 \bar{\Psi}}{\partial \bar{y}^3}$$

$$A^{\gamma_1 + \gamma_2 - \kappa_1 - \kappa_2} \frac{\partial \bar{\Psi}}{\partial \bar{y}} \frac{\partial \bar{\Theta}}{\partial \bar{x}} - A^{\gamma_1 + \gamma_2 - \kappa_1 - \kappa_2} \frac{\partial \bar{\Psi}}{\partial \bar{x}} \frac{\partial \bar{\Theta}}{\partial \bar{y}} = a A^{2\gamma_2 - \kappa_2} \frac{\partial^2 \bar{\Theta}}{\partial \bar{y}^2} \tag{6.118}$$

If one requires that the differential equations are completely invariant to the proposed linear transformations, the following coupled algebraic equations are obtained

$$\gamma_1 + 2\gamma_2 - 2\kappa_1 \quad = -\kappa_2 \tag{6.119}$$
$$3\gamma_2 - \kappa_1 \qquad\quad = -\kappa_2$$
$$\gamma_1 + \gamma_2 - \kappa_1 - \kappa_2 = 2\gamma_2 - \kappa_2$$

Under the assumption that the exponent $\gamma_1 \neq 0$, a ratio $\gamma = \gamma_2 / \gamma_1$ can be defined. Thus one obtains from Eq. (6.119) the following relations

$$\gamma = \frac{\gamma_2}{\gamma_1}, \quad \frac{\kappa_1}{\gamma_1} = 1 - \gamma, \quad \frac{\kappa_2}{\gamma_1} = 1 - 4\gamma \tag{6.120}$$

Introducing Eq. (6.120) into Eq. (6.116) results in

$$\eta = \frac{y}{x^\gamma}, \quad \Psi = x^{1-\gamma} F(\eta), \quad \Theta = x^{1-4\gamma} G(\eta) \tag{6.121}$$

with the yet unknown coefficient γ. However, if one introduces the expression for the temperature into the boundary conditions, Eq. (6.113), one obtains

$$\eta = 0: \quad x^{1-4\gamma} G(0) = 1 \tag{6.122}$$
$$\eta \to \infty: x^{1-4\gamma} G(\infty) = 0$$

From the boundary condition for $\eta \to \infty$ it can be seen that $G(\infty) = 0$. From the second boundary condition for $\eta = 0$ it is obvious that the exponent γ must be 1/4. Inserting this value of γ into Eqs. (6.121) yields

$$\eta = \frac{y}{x^{1/4}}, \quad \Psi = x^{3/4} F(\eta), \quad \Theta = G(\eta) \tag{6.123}$$

Transforming the individual terms into Eqs. (6.111-6.113) results in

$$\left(\frac{\partial \Psi}{\partial y}\right)_x = x^{1/2} F'(\eta), \quad \left(\frac{\partial \Psi}{\partial x}\right)_y = \frac{3}{4} x^{-1/4} F(\eta) - \frac{1}{4} x^{-1/4} \eta F'(\eta), \tag{6.124}$$

$$\left(\frac{\partial^2 \Psi}{\partial y^2}\right)_x = x^{1/4} F''(\eta), \quad \left(\frac{\partial^2 \Psi}{\partial x \partial y}\right) = \frac{1}{2} x^{-1/2} F'(\eta) - \frac{1}{4} x^{-1/2} \eta F'(\eta),$$

$$\left(\frac{\partial^3 \Psi}{\partial y^3}\right)_x = F'''(\eta)$$

where a prime in Eq. (6.124) indicates a differentiation with respect to the similarity coordinate η. Inserting these expressions into Eqs. (6.111-6.113) results with $\nu = \mu / \rho$ in

$$-\frac{1}{2} \left(F'(\eta)\right)^2 + \frac{3}{4} F(\eta) F''(\eta) + \nu F'''(\eta) + g \beta (T_W - T_\infty) G(\eta) = 0 \tag{6.125}$$

$$\frac{3}{4} F(\eta) G'(\eta) + a G''(\eta) = 0 \tag{6.126}$$

with the boundary conditions

$$\eta = 0: \quad F(0) = 0, \; F'(0) = 0, \; G(0) = 1 \tag{6.127}$$
$$\eta \to \infty : F'(\infty) = 0, \qquad G(\infty) = 0$$

Introducing finally the following dimensionless quantities

$$\tilde{\eta} = \frac{y}{x^{1/4}}\left(\frac{g\beta(T_W - T_\infty)}{v^2}\right)^{1/4} = \eta \left(\frac{g\beta(T_W - T_\infty)}{v^2}\right)^{1/4} \tag{6.128}$$

$$F = \left(g\beta(T_W - T_\infty)v^2\right)^{1/4} \tilde{F}$$

into Eqs. (6.125-6.127) gives

$$-\frac{1}{2}\left(\tilde{F}'(\tilde{\eta})\right)^2 + \frac{3}{4}\tilde{F}(\tilde{\eta})\tilde{F}''(\tilde{\eta}) + \tilde{F}'''(\tilde{\eta}) + \tilde{G}(\tilde{\eta}) = 0 \tag{6.129}$$

$$\frac{3}{4}\Pr \tilde{F}(\tilde{\eta})G'(\tilde{\eta}) + G''(\tilde{\eta}) = 0 \tag{6.130}$$

$$\tilde{\eta} = 0: \quad \tilde{F}(0) = 0, \; \tilde{F}'(0) = 0, \; \tilde{G}(0) = 1 \tag{6.131}$$
$$\tilde{\eta} \to \infty : \tilde{F}'(\infty) = 0, \qquad \tilde{G}(\infty) = 0$$

The Method of the Free Variable

Let us now investigate the same problem, Eqs. (6.111-6.113), with the method of the free variable. Therefore, we introduce the new coordinates, defined by

$$\xi = x \tag{6.132}$$
$$\eta = y\,g(x)$$

In addition, we assume that the stream function can be prescribed by the following expression

$$\Psi = H(\xi)F(\eta) \tag{6.133}$$

Introducing the new coordinates Eq. (6.132) and the stream function, given by Eq. (6.133), into Eqs. (6.111-6.113) results in

$$\left(F'\right)^2 \left(H'g(\xi) + Hg'(\xi)\right)Hg(\xi) - FF''g^2(\xi)HH' = \tag{6.134}$$
$$vg^3(\xi)HF''' + g\beta(T_W - T_\infty)\Theta$$

$$a\Theta'' + F\left(\frac{H'}{g(\xi)}\right)\Theta' = 0 \tag{6.135}$$

with the boundary conditions

$$\eta = 0: \quad F = 0, F' = 0, \Theta = 1 \tag{6.136}$$
$$\eta \to \infty : F' = 0, \qquad \Theta = 0$$

In order to reduce Eqs. (6.134-6.135) to a set of ordinary differential equations, the dependence on the coordinate ξ has to drop out from these equations. This leads to the fact that the following ordinary differential equations have to be satisfied

$$Hg\left(H'g + Hg'\right) = C_1 \tag{6.137}$$
$$g^2 HH' \qquad = C_2$$
$$g^3 H \qquad = C_3$$
$$\frac{H'}{g} \qquad = C_4$$

where C_1, C_2, C_3 and C_4 are constants. The four differential equations, given by Eq. (6.137) have to be linearly dependent on each other so that the two functions H and g can be determined uniquely. If we divide, for example, the second relation in Eq. (6.137) by the third one, the fourth relation is obtained. If we eliminate H and H' in the second relation using the third and the fourth relations, one obtains

$$C_3 C_4 = C_2 \tag{6.138}$$

Introducing the third and the fourth relations into the first one results in the following ordinary differential equation for the unknown function g

$$\frac{g'}{g^5} = \left(\frac{C_1}{C_3^2} - \frac{C_4}{C_3}\right) \tag{6.139}$$

This ordinary differential equation has the solution

$$g = \left(B_1 \xi + B_2\right)^{-1/4} \tag{6.140}$$

where B_1 and B_2 are constants. The constant B_2 can be set to zero, because it only shifts the origin of the coordinate. For the function H one obtains from Eq. (6.137)

$$H = B_3 \xi^{3/4} \tag{6.141}$$

Introducing Eqs. (6.140-6.141) into the Eqs. (6.132-6.133), one obtains for the similarity coordinate and for the stream function

$$\eta = \frac{y}{x^{1/4}}, \quad \Psi = x^{3/4} F(\eta) \tag{6.142}$$

This are the same expressions which have been derived by using the group-theory method (see Eq. (6.123)).

6.4.3 Similarity Solutions of the Compressible Boundary Layer Equations

As a final example, we want to consider the compressible boundary layer equations for laminar, forced convection flow over a flat plate. Similarity solutions for compressible flows have been investigated for example by Li and Nagamatsu (1953, 1955). A good overview on this subject is provided for example by Schlichting (1982). For a compressible, laminar flow, the boundary layer equations take the form (Kays and Crawford (1993))

$$\frac{\partial}{\partial x}(\rho u) + \frac{\partial}{\partial y}(\rho v) = 0 \tag{6.63}$$

$$\rho u \frac{\partial u}{\partial x} + \rho v \frac{\partial u}{\partial y} = -\frac{dp}{dx} + \frac{\partial}{\partial y}\left(\mu \frac{\partial u}{\partial y}\right) \tag{6.143}$$

$$\rho c_p \left(u \frac{\partial T}{\partial x} + v \frac{\partial T}{\partial y} \right) = u \frac{dp}{dx} + \frac{\partial}{\partial y}\left(k \frac{\partial T}{\partial y} \right) + \mu \left(\frac{\partial u}{\partial y} \right)^2 \tag{6.144}$$

As it can be seen from the system of Eqs. (6.63, 6.143-6.144), they are strongly coupled by the density. In addition, the fluid properties may no longer be considered constant, because of the high velocities under consideration. It can be assumed that the dynamic viscosity can be approximated by

$$\frac{\mu(T)}{\mu(T_{ref})} = \left(\frac{T}{T_{ref}} \right)^\omega, \quad 0.5 \leq \omega \leq 1 \tag{6.145}$$

where T_{ref} is a reference temperature. The density is related to the pressure and the temperature by a thermal state equation and we assume that the fluid under consideration can be considered as an ideal gas.

$$\rho = \frac{p}{RT} \tag{6.146}$$

In addition, we assume that c_p is constant.

For the compressible flow, it is preferable to introduce the total enthalpy into the boundary layer equations. The total enthalpy is defined for a boundary layer flow ($v \ll u$) by

$$h_0 = c_p T + \frac{1}{2}\left(u^2 + v^2\right) = c_p T + \frac{1}{2}u^2 \qquad (6.147)$$

Introducing the total enthalpy into the energy equation (6.144) results in

$$\rho\left(u\frac{\partial h_0}{\partial x} + v\frac{\partial h_0}{\partial y}\right) = \frac{\partial}{\partial y}\left(\mu\frac{Pr-1}{Pr}u\frac{\partial u}{\partial y}\right) + \frac{\partial}{\partial y}\left(\frac{\mu}{Pr}\frac{\partial h_0}{\partial y}\right) \qquad (6.148)$$

The set of partial differential equations (6.63, 6.143, 6.148) has to be solved together with Eqs. (6.145-6.146) and the following boundary conditions

$$x = 0 : u, v, h_0 \qquad \text{given} \qquad (6.149)$$
$$y = 0 : u = v = 0, \qquad h_0 = h_W$$
$$y \to \infty : u = u_\infty(x), \qquad h_0 = h_{0\infty}$$

It is convenient to introduce a stream function into the boundary layer equations defined by

$$\rho u = \frac{\partial \Psi}{\partial y} \quad , \quad \rho v = -\frac{\partial \Psi}{\partial x} \qquad (6.150)$$

In addition, the following new coordinates are introduced

$$\zeta = \rho_\infty u_\infty \int_0^y \frac{\rho}{\rho_\infty} dy \qquad (6.151)$$

$$\xi = \int_0^x \rho_\infty u_\infty \eta_\infty d\bar{x}$$

The introduction of the coordinate transformation results in the fact that the partial differential equations, which describe the momentum and energy transport in the boundary layer, take a similar structure as those for an incompressible fluid (see Schlichting (1982), Loitsianki (1967)). The change of coordinates given by Eq. (6.151) is known in literature as the Illingworth-Stewardson transformation.

After introducing the stream function and carrying out the coordinate transformation, one obtains the following two partial differential equations from Eq. (6.143) and Eq. (6.148)

$$\frac{1}{u_\infty}\frac{du_\infty}{d\xi}\left(\frac{\partial \Psi}{\partial \zeta}\right)^2 + \frac{\partial \Psi}{\partial \zeta}\frac{\partial^2 \Psi}{\partial \xi \partial \zeta} - \frac{\partial \Psi}{\partial \xi}\frac{\partial^2 \Psi}{\partial \zeta^2} = \frac{\rho_\infty}{\rho}\frac{1}{u_\infty}\frac{du_\infty}{d\xi} + \frac{\partial}{\partial \zeta}\left[C\frac{\partial^2 \Psi}{\partial \zeta^2}\right] \qquad (6.152)$$

$$\frac{\partial \Psi}{\partial \zeta}\frac{\partial \phi}{\partial \xi} - \frac{\partial \Psi}{\partial \xi}\frac{\partial \phi}{\partial \zeta} = \frac{u_\infty^2}{h_{0\infty}}\frac{\partial}{\partial \zeta}\left(\frac{Pr-1}{Pr}C\frac{\partial \Psi}{\partial \zeta}\frac{\partial^2 \Psi}{\partial \zeta^2}\right) + \frac{\partial}{\partial \zeta}\left(\frac{C}{Pr}\frac{\partial \phi}{\partial \zeta}\right) \qquad (6.153)$$

where the following abbreviations have been used

$$C = \frac{\rho\mu}{\rho_\infty\mu_\infty}, \quad \Pr = \frac{\mu c_p}{k}, \quad \phi = \frac{h_0}{h_{0\infty}} \tag{6.154}$$

The quantity C, which appears in the above equations, is called the Chapman-Rubesin parameter. This parameter can be described as a function of temperature by inserting Eqs. (6.145-6.146) into Eq. (6.154). One obtains ($p = p_\infty$)

$$C = \frac{p}{RT}\frac{RT_\infty}{p_\infty}\left(\frac{T}{T_\infty}\right)^\omega = \left(\frac{T}{T_\infty}\right)^{\omega-1}, \quad 0.5 \le \omega \le 1 \tag{6.155}$$

The Eqs. (6.152-6.153) have to be solved together with the following boundary conditions

$$\xi = 0 : \Psi, \phi \qquad\qquad \text{given} \tag{6.156}$$

$$\zeta = 0 : \frac{\partial\Psi}{\partial\xi} = 0, \frac{\partial\Psi}{\partial\zeta} = 0, \quad \phi = \phi_W$$

$$\zeta \to \infty : \qquad \frac{\partial\Psi}{\partial\zeta} = 1, \quad \phi = 1$$

After having expressed the above equations in a suitable form, the question whether the above given set of partial differential equations has similar solutions can be addressed. This can be done using the method of the free variable. Therefore, we are going to introduce the new variables

$$\bar{x} = \xi \tag{6.157}$$
$$\eta = \zeta\, w_1(\xi)$$

into the Eqs. (6.152-6.153) and into the boundary conditions, Eq. (6.156). The function $w_1(\xi)$ is unknown and is determined during the solution process.

Performing the change of coordinates results in

$$\frac{1}{u_\infty}\frac{du_\infty}{d\bar{x}}\left(\frac{\partial\Psi}{\partial\eta}\right)^2 + \frac{w_1'}{w_1}\left(\frac{\partial\Psi}{\partial\eta}\right)^2 + \frac{\partial\Psi}{\partial\eta}\frac{\partial^2\Psi}{\partial\bar{x}\partial\eta} - \frac{\partial\Psi}{\partial\bar{x}}\frac{\partial^2\Psi}{\partial\eta^2} = \frac{\rho_\infty}{\rho}\frac{1}{w_1^2}\frac{1}{u_\infty}\frac{du_\infty}{d\bar{x}} \tag{6.158}$$
$$+ w_1\frac{\partial}{\partial\eta}\left(C\frac{\partial^2\Psi}{\partial\eta^2}\right)$$

$$\frac{\partial\Psi}{\partial\eta}\frac{\partial\phi}{\partial\bar{x}} - \frac{\partial\Psi}{\partial\bar{x}}\frac{\partial\phi}{\partial\eta} = w_1^3\frac{u_\infty^2}{h_{0\infty}}\frac{\partial}{\partial\eta}\left(\frac{\Pr-1}{\Pr}C\frac{\partial\Psi}{\partial\eta}\frac{\partial^2\Psi}{\partial\eta^2}\right) + w_1\frac{\partial}{\partial\eta}\left(\frac{C}{\Pr}\frac{\partial\phi}{\partial\eta}\right) \tag{6.159}$$

with the boundary conditions

$$\bar{x} = 0 : \Psi, g \qquad\qquad \text{given} \qquad\qquad (6.160)$$

$$\eta = 0 : \frac{\partial \Psi}{\partial \eta} = 0, \quad \frac{\partial \Psi}{\partial \bar{x}} = 0, \quad \phi = \phi_W$$

$$\eta \to \infty : w_1(\bar{x}) \frac{\partial \Psi}{\partial \eta} = 1, \qquad \phi = 1$$

After introducing the new coordinates, it may now be assumed that the stream function can be expressed by the following product

$$\Psi = w_2(\bar{x}) f(\eta) \qquad\qquad (6.161)$$

The expression given by Eq. (6.161) is obvious from the boundary condition for the steam-function for $\eta \to \infty$.

The enthalpy function ϕ is assumed to depend only on η. Introducing the assumed expressions for Ψ and ϕ into the boundary conditions, one obtains

$$\eta = 0 : f(0) = 0, \, f'(0) = 0 \qquad\qquad (6.162)$$

$$\phi(0) = \phi_W$$

$$\eta \to \infty : w_1(\bar{x}) w_2(\bar{x}) f'(\infty) = 1$$

$$\phi(\infty) = 1$$

From Eq. (6.162) one obtains immediately that $w_1(\bar{x}) w_2(\bar{x})$ has to be a constant in order to satisfy the boundary condition for $\eta \to \infty$. This can be achieved, if we set for example

$$w_2(\bar{x}) = \frac{1}{w_1(\bar{x})} \qquad\qquad (6.163)$$

which leads to the following expression for the stream function

$$\Psi = \frac{f(\eta)}{w_1(\bar{x})} \qquad\qquad (6.164)$$

Introducing the expressions for the stream function and the temperature into the Eqs. (6.158-6.159) results in the following differential equations

$$\frac{1}{w_1^2} \frac{1}{u_\infty} \frac{du_\infty}{d\bar{x}} (f')^2 + \frac{w_1'}{w_1^3} ff'' = \frac{\rho_\infty}{\rho} \frac{1}{w_1^2} \frac{1}{u_\infty} \frac{du_\infty}{d\bar{x}} + (Cf'')' \qquad\qquad (6.165)$$

$$\frac{w_1'}{w_1^3} f\phi' = \frac{u_\infty^2}{h_{0\infty}} \left(\frac{\mathrm{Pr}-1}{\mathrm{Pr}} Cf f'' \right)' + \left(\frac{C}{\mathrm{Pr}} \phi' \right)' \qquad\qquad (6.166)$$

In order to obtain similarity solutions from these two equations, the dependence on $\bar{x}$ in these equations must cancel out. This is only possible, if

$$\frac{w_1'}{w_1^3} = \text{const.} \tag{6.167}$$

Equation (6.167) can simply be integrated and one result of this equation is

$$w_1 = \frac{1}{\sqrt{2\bar{x}}} \tag{6.168}$$

Introducing Eq. (6.168) into the Eqs. (6.165-6.166) results in

$$\left(Cf''\right)' + \frac{2\bar{x}}{u_\infty}\frac{du_\infty}{d\bar{x}}\left(\frac{\rho_\infty}{\rho} - \left(f'\right)^2\right) + ff'' = 0 \tag{6.169}$$

$$\left(\frac{C}{\text{Pr}}\phi'\right)' + f\phi' + \frac{u_\infty^2}{h_{0\infty}}\left(\frac{\text{Pr}-1}{\text{Pr}}C f'f''\right) = 0 \tag{6.170}$$

These two coupled differential equations have to be solved together with the following boundary conditions

$$\eta = 0 : f(0) = 0, \; f'(0) = 0, \; \phi(0) = \phi_w \tag{6.171}$$

$$\eta \to \infty : \qquad f'(\infty) = 1, \; \phi(\infty) = 1$$

In order to guarantee that the Eqs. (6.169-6.170) are a set of ordinary differential equations, any dependence on $\bar{x}$ must cancel out from these equations. This fact leads to the following restrictions on the coefficients in these equations:

1.   The Prandtl number must be a constant or has to depend only on η. This is not a strong restriction, because, for example for air, the Prandtl number is approximately a constant for a large temperature range.
2.   The Chapman-Rubesin parameter must be a constant or a function of η only. This is no real restriction, as can be seen from Eq. (6.155). For the viscosity law considered here, this is always satisfied.
3.   The density ratio ρ_∞ / ρ has to be constant or a function of η only. Because we considered an ideal gas, we obtain from Eq. (6.146) that

$$\frac{\rho_\infty}{\rho} = \frac{T}{T_\infty} \tag{6.172}$$

4.   The expression

$$\frac{2\bar{x}}{u_\infty}\frac{du_\infty}{d\bar{x}} = \beta = \text{const.} \tag{6.173}$$

has to be constant. This is possible if the external flow speed $u_\infty = \text{const.}$, or in general if $u_\infty = C_1 \bar{x}^m$.

5. Finally, one has to require that the expression

$$\frac{u_\infty^2}{h_{0\infty}} = 2\left(1 + \frac{2}{(\kappa - 1)\,\mathrm{Ma}_\infty^2}\right)^{-1} \tag{6.174}$$

has to be constant, or the Prandtl number must be equal to one. For most gases the assumption that $\mathrm{Pr} = 1$ is quite reasonable. If this is not the case, it is required that either the Mach number is constant or that the Mach number is high enough, so that $u_\infty^2 / h_{0\infty}$ can be approximated by 2.

The similarity equations (6.169-6.170) have first been obtained by Cohen and Reshotko (1956) (see also Eckert and Drake (1987) and Jischa (1982)).

Flow and Heat Transfer on a Flat Plate

As an illustration of the preceding analysis we consider a flow over a flat plate. The free stream velocity $u_\infty = \text{const.}$ and the dynamic viscosity is described by

$$\frac{\mu(T)}{\mu(T_{ref})} = \frac{T}{T_{ref}} \tag{6.175}$$

This leads to the fact that the Chapman-Rubesin parameter $C = 1$ and the Eqs. (6.169-6.170) simplify to

$$f''' + ff'' = 0 \tag{6.176}$$

$$K\left(f' f''\right)' + \vartheta'' + \mathrm{Pr}\, f\vartheta' = 0 \tag{6.177}$$

with the abbreviations

$$\vartheta = \frac{\phi - \phi_W}{1 - \phi_W} \ , \quad K = \frac{2(\mathrm{Pr} - 1)}{1 - \phi_W}\left(1 + \frac{2}{(\kappa - 1)\,\mathrm{Ma}_\infty^2}\right)^{-1} \tag{6.178}$$

The Eqs. (6.176-6.177) have to be integrated with the boundary conditions

$$\eta = 0: \quad f(0) = 0,\ f'(0) = 0,\ \vartheta(0) = 0 \tag{6.179}$$

$$\eta \to \infty: f'(\infty) = 1, \qquad \vartheta(\infty) = 1$$

Under these assumptions, the momentum equation reduces to the Blasius equation. (see Prandtl (1935) and Schlichting (1982)). The energy equation can be solved by standard methods and the final result is given by

$$\vartheta = 1 - \frac{\int\limits_{\eta}^{\infty}\left(f''\right)^{\mathrm{Pr}}d\bar\eta}{\int\limits_{0}^{\infty}\left(f''\right)^{\mathrm{Pr}}d\bar\eta} + K\int\limits_{0}^{\infty}\left(f''\right)^{\mathrm{Pr}}\int\limits_{0}^{t}\left(f''\right)^{-\mathrm{Pr}}\left(f'f''\right)' d\bar\eta\, dt \qquad (6.180)$$

$$\cdot\left\{\frac{\left[\int\limits_{\eta}^{\infty}\left(f''\right)^{\mathrm{Pr}}d\bar\eta\quad\int\limits_{\eta}^{\infty}\left(f''\right)^{\mathrm{Pr}}\int\limits_{0}^{t}\left(f''\right)^{-\mathrm{Pr}}\left(f'f''\right)' d\bar\eta\, dt\right]}{\left[\int\limits_{0}^{\infty}\left(f''\right)^{\mathrm{Pr}}d\bar\eta\quad\int\limits_{0}^{\infty}\left(f''\right)^{\mathrm{Pr}}\int\limits_{0}^{t}\left(f''\right)^{-\mathrm{Pr}}\left(f'f''\right)' d\bar\eta\, dt\right]}\right\}$$

After the temperature distribution is known for this case, the Nusselt number distribution can be predicted. The Nusselt number is defined by

$$\mathrm{Nu}_x = \frac{-x\dfrac{\partial T}{\partial y}\bigg|_{y=0}}{T_{0\infty}-T_W} \qquad (6.181)$$

Introducing the new coordinates into the Nusselt number results in

$$\mathrm{Nu}_x = \sqrt{\frac{\mathrm{Re}_x}{2}}\left(\frac{\rho_W}{\rho_\infty}\right)\vartheta'(0) \qquad (6.182)$$

with the dimensionless enthalpy gradient at the wall

$$\vartheta'(0) = \frac{\left(f''(0)\right)^{\mathrm{Pr}}}{\int\limits_{0}^{\infty}\left(f''\right)^{\mathrm{Pr}}d\bar\eta}\left\{1+K\int\limits_{0}^{\infty}\left(f''\right)^{\mathrm{Pr}}\int\limits_{0}^{t}\left(f''\right)^{-\mathrm{Pr}}\left(f'f''\right)d\bar\eta\, dt\right\} \qquad (6.183)$$

Problems

6-1. Consider the heat conduction in a rectangular flat plate. The plate has a heat conductivity which depends only on temperature. The problem can be described by the following equation

$$\frac{\partial}{\partial x}\left(k(T)\frac{\partial T}{\partial x}\right)+\frac{\partial}{\partial y}\left(k(T)\frac{\partial T}{\partial y}\right)=0$$

where the heat conductivity is assumed to be given by the following equation

$$k(T) = k_0\frac{T}{T_0}$$

where k_0, T_0 are constant reference values.

a.)   Investigate, first, the case of a constant heat conductivity $k = k_0$. Show by insertion, that if T_1 and T_2 are both solutions of the heat conduction equation, then also $C_1 T_1 + C_2 T_2$ is a solution of the equation (C_1 and C_2 are arbitrary constants).

b.)   Consider now the case of a temperature dependent heat conductivity. Show that if two solutions of the problem have been found (T_1 and T_2), the sum of them $T_1 + T_2$ is not a solution of the problem.

6-2. Consider the one dimensional transient heat conduction in a slab ($0 \leq x \leq L$). The slab is initially at a constant temperature T_0. For $t > 0$ the boundary condition are: at $x = 0$, the slab is insulated and at $x = L$, it is kept at a constant temperature T_W. The problem under consideration is described by the following equation

$$\rho c \frac{\partial T}{\partial t} = \frac{\partial}{\partial x}\left(k \frac{\partial T}{\partial x}\right)$$

with the boundary conditions

$$t = 0: \ T = T_0$$
$$x = 0: \frac{\partial T}{\partial x} = 0$$
$$x = L: T = T_W$$

The heat conductivity of the solid is assumed to depend on temperature in the form

$$k = k_0(1 + C(T - T_0))$$

However, the heat diffusivity $a = k/(\rho c)$ can be taken as constant.

a.)   Apply the Kirchhoff transformation to the above equations and show that the resulting heat conduction equation reduces to a linear equation if the heat diffusivity is assumed constant.

b.)   Solve the transformed problem by using the method of separation of variables.

c.)   What is the solution of the original problem?

6-3. Consider the partial differential equation

$$\tilde{y} \frac{\partial \Theta}{\partial \tilde{x}} = \frac{\partial^2 \Theta}{\partial \tilde{y}^2}$$

with the boundary conditions

$$\tilde{x} = 0: \quad \Theta = 1$$
$$\tilde{y} = 0: \quad \Theta = 0$$
$$\tilde{y} \to \infty: \Theta = 1$$

For this problem, similarity solutions can be obtained with the method of the free variable.

a.) Introduce the new variables

$$\xi = \tilde{x}$$
$$\eta = \tilde{y}\, g(\tilde{x})$$
$$\Theta = \Theta(\eta)$$

into the differential equation. What is the resulting equation (please remember that Θ depends only on η)?

b.) Which ordinary differential equation has to be satisfied for the function $g(\tilde{x})$, so that similarity solutions are possible?

c.) Solve the ordinary differential equation for g and determine the similarity coordinate.

d.) Determine the solution for the above given problem.

6-4. Consider the transient heat conduction in a semi-infinite body. The body contains a heat source. At the surface of the body, a time dependent temperature is applied. The process can be described by the following nonlinear partial differential equation

$$\rho c \frac{\partial T}{\partial t} = k \frac{\partial^2 T}{\partial x^2} + F(T,t)e^{-x/\sqrt{t}}$$

with boundary conditions

$$x = 0: \quad T = T_W(t)$$
$$x \to \infty: T = T_0$$
$$t = 0: \quad T = T_0$$

All properties (ρ, c, k) of the material can be considered constant. Investigate under which circumstances similarity solutions of the above equation are possible, assuming the similarity variable is given by $\eta = C x/\sqrt{t}$, C=const. .

a.) Introduce the dimensionless temperature $\Theta = (T - T_0)/(T_W(t) - T_0)$ into the above equation and boundary conditions.

b.) Determine the constant C in the similarity variable from dimensional reasoning, so that η is dimensionless.

c.) Consider now the case $F(T,t)=0$. Perform the coordinate transformation and show that similarity solutions are possible for the boundary conditions under consideration.

d.) Consider now the case $T_W = \text{const.}$. Determine for this case the form of the function $F(T,t)$, for which similarity solutions are possible employing the similarity variable suggested above.

6-5. Consider the flow over a flat plate with an external pressure gradient. The flow is incompressible and all fluid properties are constant. The problem is described by the incompressible boundary layer equations given by

$$\frac{\partial u}{\partial x} + \frac{\partial v}{\partial y} = 0$$

$$u\frac{\partial u}{\partial x} + v\frac{\partial u}{\partial y} = -\frac{1}{\rho}\frac{dp}{dx} + \frac{\mu}{\rho}\frac{\partial^2 u}{\partial y^2}$$

with boundary conditions

$$y = 0: \ u = v = 0$$
$$y \rightarrow \infty : u = u_\infty(x)$$

The pressure gradient in the momentum equation is linked to the free-stream acceleration by

$$-\frac{1}{\rho}\frac{dp}{dx} = u_\infty \frac{du_\infty}{dx}$$

a.) Introduce a stream function into the momentum equation, so that the continuity equation is automatically satisfied. Express the boundary conditions in terms of the stream function. What are the resulting equations?

b.) Investigate the equations obtained above by using the group-theory method. What is the similarity coordinate and how has the stream function to look like so that similarity solutions are possible? What ordinary differential equation can be obtained for the modified stream function?

c.) Investigate the problem under b.) once again using the method of the free variable. Can you verify the results obtained under b.)?

6-6. Consider the one dimensional transient heat conduction in a semi-infinite solid. The temperature distribution can be described by the following equations

$$\frac{\partial T}{\partial t} = a\frac{\partial}{\partial x}\left[\left(1 + b\left(\frac{T - T_0}{T_W - T_0}\right)\frac{\partial T}{\partial x}\right)\right]$$

$$t = 0: \quad T = T_0$$
$$x = 0: \quad T = T_W$$
$$x \rightarrow \infty : T = T_0$$

where the heat diffusivity a can be considered to be constant. The temperatures T_0 and T_W and the value b are constant.

a.) Introduce a suitable dimensionless temperature into the above equations.

b.) For this problem, similarity solutions can be obtained using the method of the free variable. Introduce the new coordinate $\eta = x^2 g(t)$ into the heat conduction equation under the assumption that the dimensionless temperature only depends on the similarity coordinate. What ordinary differential equation has to be satisfied for the function $g(t)$? What similarity coordinate is obtained? Determine the similarity solution of the problem.

Appendix A: The Fully Developed Velocity Profile for Turbulent Duct Flows

This appendix discusses the hydrodynamically fully developed velocity profile for pipe and channel flows. The geometry under consideration is shown in Fig. A.1

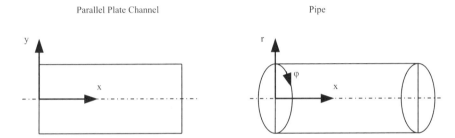

Fig. A.1: Geometry and coordinate system

For the flow in a parallel plate channel, the velocity components in the x, y and z direction are u, v and w, whereas for the flow in a circular pipe, the velocity components u, v and w denote the flow in the x, r and φ direction.

We restrict our considerations to an incompressible flow with constant fluid properties. The Reynolds averaged Navier-Stokes equations and the continuity equation are given for a flow in a parallel plate channel by (Kays and Crawford (1993))

Parallel Plate Channel

$$\rho\left(u\frac{\partial u}{\partial x}+v\frac{\partial u}{\partial y}\right)=-\frac{\partial p}{\partial x}+\mu\left(\frac{\partial^2 u}{\partial x^2}+\frac{\partial^2 u}{\partial y^2}\right)-\rho\frac{\partial\overline{(u'u')}}{\partial x}-\rho\frac{\partial\overline{(u'v')}}{\partial y} \tag{A.1}$$

$$\rho\left(u\frac{\partial v}{\partial x}+v\frac{\partial v}{\partial y}\right)=-\frac{\partial p}{\partial y}+\mu\left(\frac{\partial^2 v}{\partial x^2}+\frac{\partial^2 v}{\partial y^2}\right)-\rho\frac{\partial\overline{(u'v')}}{\partial x}-\rho\frac{\partial\overline{(v'v')}}{\partial y} \tag{A.2}$$

$$\frac{\partial u}{\partial x}+\frac{\partial v}{\partial y}=0 \tag{A.3}$$

For the flow in a circular pipe with rotational symmetry, the equations are given by (Kays and Crawford (1993))

Pipe

$$\rho\left(u\frac{\partial u}{\partial x}+v\frac{\partial u}{\partial r}\right)=-\frac{\partial p}{\partial x}+\frac{\mu}{r}\frac{\partial}{\partial r}\left(r\frac{\partial u}{\partial r}\right)+\mu\frac{\partial^2 u}{\partial x^2} \tag{A.4}$$
$$-\frac{\rho}{r}\frac{\partial}{\partial r}\left(r\overline{u'v'}\right)-\rho\frac{\partial}{\partial x}\left(\overline{u'u'}\right)$$

$$\rho\left(u\frac{\partial v}{\partial x}+v\frac{\partial v}{\partial r}\right)=-\frac{\partial p}{\partial r}+\frac{\mu}{r}\frac{\partial}{\partial r}\left(r\frac{\partial v}{\partial r}\right)+\mu\frac{\partial^2 v}{\partial x^2} \tag{A.5}$$
$$-\frac{\rho}{r}\frac{\partial}{\partial r}\left(r\overline{v'v'}\right)-\rho\frac{\partial}{\partial x}\left(\overline{u'v'}\right)-\rho\frac{\overline{w'w'}}{r}$$

$$\frac{\partial u}{\partial x}+\frac{1}{r}\frac{\partial}{\partial r}(rv)=0 \tag{A.6}$$

For a hydrodynamically fully developed flow, the axial velocity component does not change (by definition) with the axial position and therefore $\partial u/\partial x$ is zero. If we introduce this result into the continuity equation (A.3) or (A.6), we obtain that the radial velocity component v is equal to zero and the continuity equation is then automatically satisfied. The fact that $v = 0$ for the fully developed flow is obvious, as a non-zero velocity component v would lead automatically to a change in the axial velocity component for different axial positions. Introducing these results into the above given equations results in

Parallel Plate Channel

$$0=-\frac{\partial p}{\partial x}+\mu\frac{d^2 u}{dy^2}-\rho\frac{d(\overline{u'v'})}{dy} \tag{A.7}$$

$$0=-\frac{\partial p}{\partial y}-\rho\frac{d\left(\overline{v'v'}\right)}{dy} \tag{A.8}$$

Pipe

$$0=-\frac{\partial p}{\partial x}+\frac{\mu}{r}\frac{d}{dr}\left(r\frac{du}{dr}\right)-\frac{\rho}{r}\frac{d}{dr}\left(r\overline{u'v'}\right) \tag{A.9}$$

$$0 = -\frac{\partial p}{\partial r} - \frac{\rho}{r}\frac{d}{dr}\left(r\overline{v'v'}\right) - \rho\frac{\overline{w'w'}}{r} \tag{A.10}$$

The above equations have to be solved together with the following boundary conditions:

Parallel Plate Channel

$$y = 0 : \frac{du}{dy} = 0, \tag{A.11}$$

$$y = h : u = 0$$

Pipe

$$r = 0 : \frac{du}{dr} = 0, \tag{A.12}$$

$$r = R : u = 0$$

In the above equations, the partial derivatives for the axial velocity component and the turbulent stresses have been replaced by total derivatives, because the functions are only dependent on the coordinate orthogonal to the flow direction.

The equations describing the hydrodynamically fully developed flow in the parallel plate channel and in the pipe are quite similar, as it can be seen by comparing Eq. (A.7) with (A.9) and Eq. (A.8) with (A.10). They can be written in a condensed form by using a flow superscript F. For the flow in a parallel plate channel $F = 0$, whereas for the flow in a pipe $F = 1$. This results in the following equations:

$$0 = -\frac{\partial p}{\partial x} + \frac{1}{r^F}\frac{d}{dn}\left[r^F\left(\mu\frac{du}{dn} - \rho\overline{u'v'}\right)\right] \tag{A.13}$$

$$0 = \rho\left[\frac{d\overline{(v')^2}}{dn} + F\left(\frac{\overline{(v')^2} - \overline{(w')^2}}{r}\right)\right] + \frac{\partial p}{\partial n} \tag{A.14}$$

with the boundary conditions

$$n = 0 : \frac{du}{dn} = 0, \tag{A.15}$$

$$n = L : u = 0$$

In these equations, n denotes the coordinate orthogonal to the flow direction ($n = r$ for the flow in a pipe and $n = y$ for the flow in a parallel plate channel). The quantity $L = R$ (pipe radius) for pipe flow and $L = h$ (half channel height) for the flow in a parallel plate channel.

The Eqs. (A.13-A.14) are solved together with the boundary conditions given by Eq. (A.15). Eq. (A.14) can directly be integrated and results in

$$p = p_W(x) - \rho \overline{(v')^2} + \rho F \int_r^R \frac{\overline{(v')^2} - \overline{(w')^2}}{\bar{r}} \, d\bar{r} \qquad (A.16)$$

Equation (A.16) shows nicely that the pressure in the direction orthogonal to the main flow changes across the duct due to the presence of the turbulent stresses. For a parallel plate channel, the change in pressure is only caused by the normal turbulent stress, whereas for a pipe flow also the difference between the normal turbulent stresses in radial and circumferential direction causes an additional change. Differentiating Eq. (A.16) with respect to x leads to

$$\frac{\partial p}{\partial x} = \frac{\partial p_W}{\partial x}$$

because the turbulent stresses only depend on the coordinate orthogonal to the flow direction.

For the flows considered here through a pipe or a parallel plate channel, the mass flow rate is constant and the integral form of the mass continuity equation leads to

$$\bar{u} \, L^{(F+1)} = (F+1) \int_0^L u \, r^F \, dn \qquad (A.17)$$

where $\bar{u}$ denotes the mean axial velocity in the duct. The above given equation determines the yet unknown pressure gradient in the axial direction ($\partial p / \partial x$ is a constant for the hydrodynamically fully developed flow). In order to solve Eq. (A.13), the turbulent shear stress $-\rho \overline{u'v'}$ has to be related to the mean flow field. This can be done using a simple mixing length model. The use of the simple mixing length model is adequate for this type of flows (Cebeci and Bradshaw (1984)). However, for more complicated flows, like the flow through a pipe with cyclic constrictions, mixing length models may lead to inaccurate answers. The reader is referred to Reynolds (1974), Launder (1988) or Kays or Crawford (1993) for a review of more advanced turbulence models. Using the mixing length model, the turbulent shear stress can be written as

$$-\overline{u'v'} = \varepsilon_m \frac{du}{dn} \qquad (A.18)$$

with the eddy viscosity ε_m given by

$$\varepsilon_m = l^2 \left| \frac{du}{dn} \right| \tag{A.19}$$

The mixing length l has been measured by Nikuradse (1932) for pipe flows. He obtained the interesting result that l does not depend on the Reynolds number for $Re_D > 10^5$. The mixing length distribution can be approximated by (see Schlichting (1982))[1]

$$\frac{l}{L} = 0.14 - 0.08 \left(\frac{n}{L} \right)^2 - 0.06 \left(\frac{n}{L} \right)^4 \tag{A.20}$$

Near the wall, the mixing length distribution, according to Eq. (A.20), approaches the one given by Prandtl (see Schlichting (1982))

$$l = \kappa (L - n), \quad \kappa = 0.4 \tag{A.21}$$

Close to the wall, the turbulent fluctuations are damped out and, below a certain dimensionless distance away from the wall, the turbulent fluctuations are zero. In order to use the mixing length distribution throughout the whole calculation region, it is convenient to modify the expression using a damping function This guarantees that the mixing length tends to zero within the laminar sublayer. Cebeci and Bradshaw (1984) modified the mixing length distribution, given by Eq. (A.20), with the van Driest damping term (van Driest (1956)). This leads to

$$\frac{l}{L} = \left[0.14 - 0.08 \left(\frac{n}{L} \right)^2 - 0.06 \left(\frac{n}{L} \right)^4 \right] \left\{ 1 - \exp \left(-\frac{y^+}{26} \right) \right\} \tag{A.22}$$

with the abbreviations

$$y^+ = \frac{y_w u_\tau}{v}, \quad u_\tau = \sqrt{\frac{|\tau_w|}{\rho}}, \quad y_w = (L - n) \tag{A.23}$$

Introducing Eq. (A.18) into Eq. (A.12) results in an ordinary differential equation for the axial velocity component

$$0 = -\frac{\partial p_W}{\partial x} + \frac{1}{r^F} \frac{d}{dn} \left[r^F \left(\mu \frac{du}{dn} + \rho l^2 \frac{du}{dn} \left| \frac{du}{dn} \right| \right) \right] \tag{A.24}$$

This equation has to be solved together with the boundary conditions given by Eq. (A.14). The axial pressure gradient $\partial p_W / \partial x$ can be replaced by the shear stress ve-

---

[1] Zagustin an Zagustin (1969) developed theoretically a formula for the mixing length distribution in a pipe by solving the conservation equation for the turbulent energy. Their mixing length distribution is also in good agreement with the measurements of Nikuradse (1932).

locity u_τ. In order to do this, let us consider the forces acting on a small duct element (shown in Fig. A.2)

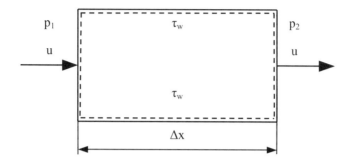

Fig. A.2: Force balance for a small duct element

Equating the forces acting on the small duct element shown in Fig. A.2 results in:

$$(p_1 - p_2)\,\pi R^2 = \tau_W\, 2\pi R\, \Delta x, \qquad \text{Pipe} \tag{A.25}$$

$$(p_1 - p_2)\, 2h = \tau_W\, 2\Delta x, \qquad \text{Parallel Plates} \tag{A.26}$$

From the above two equations, one obtains the following relation between the pressure gradient and the wall shear stress

$$-\frac{\partial p}{\partial x} = \frac{\tau_W\,(F+1)}{L} = \frac{u_\tau^2\, \rho\,(F+1)}{L} \tag{A.27}$$

Introducing this result and the following dimensionless quantities into Eq. (A.24)

$$\tilde{r} = \frac{r}{L}, \quad \tilde{n} = \frac{n}{L}, \quad u^+ = \frac{u}{u_\tau}, \quad \tilde{l} = \frac{l}{L}, \quad \mathrm{Re}_\tau = \frac{u_\tau L}{\nu} \tag{A.28}$$

results in the following ordinary differential equation for the axial velocity component

$$-\mathrm{Re}_\tau\, \tilde{r}^F\,(F+1) = \frac{d}{d\tilde{n}}\left[\tilde{r}^F\left(\frac{du^+}{d\tilde{n}} - \tilde{l}^2\,\mathrm{Re}_\tau\left(\frac{du^+}{d\tilde{n}}\right)^2\right)\right] \tag{A.29}$$

Integration of this equation and application of the boundary condition $du^+/d\tilde{n} = 0$ for $\tilde{n} = 0$ results in

$$-\mathrm{Re}_\tau\, \tilde{n} = \frac{du^+}{d\tilde{n}} - \tilde{l}^2\,\mathrm{Re}_\tau\left(\frac{du^+}{d\tilde{n}}\right)^2 \tag{A.30}$$

After solving for the unknown velocity gradient one obtains from Eq. (A.30)

$$\frac{du^+}{d\tilde{n}} = \frac{-2\tilde{n}\,\mathrm{Re}_\tau}{1+\sqrt{1+4\tilde{n}\tilde{l}^2\,\mathrm{Re}_\tau^{\,2}}} \qquad (A.31)$$

This ordinary differential equation has to be solved numerically with the boundary condition that $u^+ = 0$ for $\tilde{n}=1$. In addition, the continuity equation in integral form, Eq. (A.17), has to be satisfied. This equation reads after introducing the dimensionless quantities according to Eq. (A.28)

$$\frac{\mathrm{Re}_L}{\mathrm{Re}_\tau} = (F+1)\int_0^1 u^+ \tilde{r}^F d\tilde{n}\,, \quad \mathrm{Re}_L = \frac{\bar{u}\,L}{\nu} \qquad (A.32)$$

For a selected Reynolds number Re_L, a value for Re_τ has to be estimated. After this, Eq. (A.31) is solved numerically and the resulting velocity distribution is inserted into Eq. (A.32). From Eq. (A.32) a new value for Re_τ is obtained and the iteration is carried on up to the point, where both Eqs. (A.31-A.32) are satisfied. Fig. A.3 shows a comparison between predicted and measured velocity profiles in a planar channel and in a pipe.

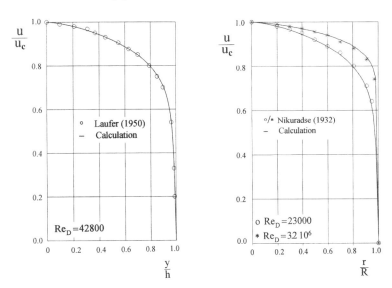

Fig. A.3: Fully developed velocity profiles in a planar channel and in a pipe

It can be seen that the application of the simple mixing length model leads to quite satisfactory results for the fully developed flow in a pipe and in a parallel plate channel.

Note that also the hydrodynamic developing flow in a duct can be predicted as well by using a modified mixing length model. The reader is referred for more details to Cebeci and Chang (1978) and Cebeci and Bradshaw (1984).

Antonia and Kim (1991) compared calculations based on a mixing length model for hydrodyamically fully developed flow in a channel with experiments and also with direct numerical simulations (DNS). They found good agreement between their calculations and the DNS data.

After establishing a simple calculation method for the hydrodynamically fully developed flow, the *universality of the velocity profiles* for turbulent pipe flows is addressed next. As it can be seen from Eq. (A.12), the total shear stress τ_{rx} for the pipe flow consists of a molecular part and of the turbulent shear stress:

$$\tau_{rx} = (\mu + \rho\,\varepsilon_m)\frac{du}{dr} \tag{A.33}$$

After integrating Eq. (A.12) it can be seen that the total shear stress τ_{rx} varies linearly across the pipe. With this result, one obtains from Eq. (A.33)

$$\frac{du^+}{dy^+} = \frac{\tau_{rx}/\tau_W}{1+\dfrac{\varepsilon_m}{v}} = \frac{1 - y_W/R}{1+\dfrac{\varepsilon_m}{v}} \tag{A.34}$$

where $y_W = R - r$. Eq. (A.34) has been the basis for many considerations for turbulent pipe flows, which led finally to approximations for the fully developed turbulent velocity distribution. In the following section, we describe mainly the work of Reichhardt (1951) on this topic.

The Law of the Wall

Near the wall, the eddy viscosity distribution ε_m can be approximated relatively easy from experimental data. Nikuradse (1932) derived from his experimental results the following approximation for ε_m

$$\frac{\varepsilon_m}{v} = \kappa\,y^+, \quad y^+ = \frac{y_W u_\tau}{v} \tag{A.35}$$

Introducing Eq. (A.35) into Eq. (A.34) and integrating the resulting expression for the region close to the wall ($y_W/R \ll 1$ and $y^+ \gg 1$) results in

$$u^+ = \frac{1}{\kappa}\ln\,y^+ + c \tag{A.36}$$

The logarithmic velocity distribution in Eq. (A.36) has been derived under various assumptions by Prandtl (1925) and von Karman (1939). It is a very important result for turbulent flows, because of the universality of this velocity distribution for various turbulent flow fields. From Eq. (A.35), it can be seen that this equation looses its meaning, if we consider values of y^+ very close to the wall, as for

$y^+ < 11$ the turbulent exchange in the viscous sublayer of the flow tends to zero. In order to overcome this difficulty, Reichardt (1951) developed the following distribution for the eddy viscosity near the wall

$$\frac{\varepsilon_m}{\nu} = \kappa\left(y^+ - y_1^+ \arctan\left(\frac{y^+}{y_1^+}\right)\right), \qquad y_1^+ = 11 \tag{A.37}$$

This function goes to zero for $y^+ \to 0$. For larger values of y^+, Eq. (A.37) approaches continuously Eq. (A.35). Introducing Eq. (A.37) into Eq. (A.34) and integrating, results after restricting again to the region close to the wall ($y_W / R \ll 1$ and $y^+ \gg 1$) in

$$u^+ = 2.5\ln(1+\kappa y^+) + 7.8\left[1 - \exp\left(-\frac{y^+}{y_1^+}\right) - \left(\frac{y^+}{y_1^+}\right)\exp\left(-\frac{y^+}{3}\right)\right] \tag{A.38}$$

Eq. (A.38) compares favorably with experimental data in the near wall region (see Reichardt (1951)).

The Core Region

In the central part of the pipe, all distributions have to be symmetric to $\tilde{r} = 0$. Therefore, it is useful to analyze this area with Eq. (A.34) after introducing the radial coordinate instead of the wall coordinate. Furthermore, the molecular viscosity can be neglected compared to the turbulent eddy viscosity in the denominator of Eq. (A.34). This leads to

$$\frac{du^+}{d\tilde{r}} = \frac{\tilde{r}}{\mathrm{Re}_\tau \dfrac{\varepsilon_m}{\nu}} \tag{A.39}$$

In the central region of the pipe, the eddy viscosity can be approximated according to Reichhardt (1951) by

$$\frac{\varepsilon_m}{\nu} = \mathrm{Re}_\tau \frac{\kappa}{3}\left(\frac{1}{2} + \tilde{r}^2\right)\left(1 - \tilde{r}^2\right) \tag{A.40}$$

Eq. (A.40) was found to be in good agreement with experimental data. If we introduce Eq. (A.40) into Eq. (A.39) and integrate, the following result for the velocity distribution is obtained

$$\frac{u_C - u}{u_\tau} = \frac{1}{\kappa}\ln\left(\frac{1+2\tilde{r}^2}{1-\tilde{r}^2}\right) \tag{A.41}$$

where u_C denotes the maximum velocity in the pipe at $\tilde{r} = 0$. Eq. (A.41) is valid throughout the flow area, except the region close to the wall, where the molecular viscosity plays an important role (Reichhardt (1951)).

An Approximate Velocity Distribution for the Whole Flow Field

Based on the two approximate velocity distributions, Eq. (A.38) and Eq. (A.41), Reichhardt (1951) developed an approximate velocity distribution, which is valid throughout the flow area.

$$u^+ = \frac{1}{\kappa} \ln \left\{ \left(1 + \kappa y^+\right) \frac{\frac{3}{2}\,(\,1 + \tilde{r}\,)}{1 + 2\,\tilde{r}^2} \right\} + C_1 \left[1 - \exp\left(-\frac{y^+}{y_1^+}\right) - \left(\frac{y^+}{y_1^+}\right)\exp\left(-\frac{y^+}{3}\right) \right] \tag{A.42}$$

with the constant $C_1 = 5.5 - 1/\kappa \ln \kappa$. Eq. (A.42) agrees very well with experimental results of Nikuradse (1932) and Reichhardt (1940).

During the past 90 years, a large number of different expressions for the universal velocity distribution in a pipe have been developed by different researchers. All equations are based on slightly different assumptions for the eddy viscosity distribution in the pipe and all the equations describe more or less accurately the fully developed velocity distribution. Table A.1 gives an overview over several different expressions for the velocity distribution in the pipe.

Strictly speaking, the above derived universal velocity distribution is only valid for large Reynolds numbers ($Re_D > 10^5$). However, it was found that they approximate very well the velocity distribution also for much lower Reynolds numbers. Rothfus et al. (1958) and Dwyer (1965) showed that the influence of the Reynolds number on the universal velocity distribution $u^+ = f(y^+)$ can be incorporated by using the following quantities

$$y^{++} = y^+ / \left(\frac{\bar{u}}{u_C}\right)_{\text{Pipe}}, \quad u^{++} = u^+ / \left(\frac{\bar{u}}{u_C}\right)_{\text{Pipe}} \tag{A.43}$$

In addition, Rothfus et al. (1958) and Dwyer (1965) were able to show that the modified velocity distribution $u^{++} = f(y^{++})$ can also be applied to the flow in a parallel channel and in a concentric circular annulus. The agreement between experimental data and the velocity distribution $u^{++} = f(y^{++})$ given by Rothfus et al. (1958) for a wide range of Reynolds numbers is good.

Table A.1: Universal velocity profiles for hydrodynamically fully developed duct flows

Reference	Functional dependence $u^+ = f(y^+)$	Validity range
Prandtl (1910)	$u^+ = y^+$	$0 \le y^+ \le 11.5$
Taylor (1916)	$u^+ = 2.5 \ln y^+ + 5.5$	$y^+ > 11.5$
	$u^+ = y^+$	$0 \le y^+ \le 5$
von Karman (1939)	$y^+ = 5 \ln y^+ - 3.05$	$5 < y^+ \le 30$
	$u^+ = 2.5 \ln y^+ + 5.5$	$y^+ > 30$
Reichhardt (1951)	$u^+ = 2.5 \ln(1 + \kappa y^+) + 7.8 \, (1 - \exp(y^+ / y_1^+)$ $- (y^+ / y_1^+) \exp(y^+ / 3))$	for all y^+
Deissler (1954)	$u^+ = \displaystyle\int_0^{y^+} \frac{d\hat{y}}{1 + n^2 u^+ \hat{y}[1 - \exp(-n^2 n^+ \hat{y})]}, \; n = 0.124$	$0 \le y^+ \le 26$
	$u^+ = 2.78 \;\; \ln y^+ \; + \; 3.8$	$y^+ > 26$
Rannie (1956)	$u^+ = 14.53 \tanh\left(y^+ / 14.53 \right)$	$0 \le y^+ \le 27.5$
	$u^+ = 2.5 \ln y^+ + 5.5$	$y^+ > 27.5$
Spalding (1961)	$y^+ = u^+ + 0.1108[\exp(0.4u^+) - 1 - 0.4u^+$ $- (0.4u^+)^2/_{2!} - (0.4u^+)^3/_{3!} - (0.4u^+)^4/_{4!}]$	for all y^+
Notter and Schleicher (1971)	$u^+ = \dfrac{1}{0.091} \tan^{-1}\left(0.091 y^+\right)$	$0 \le y^+ \le 45$
	$u^+ = 5.1 + 2.5 \ln y^+$	$45 \le y^+ < \bar{y}_W$ $\bar{y}_W = 0.15$
	$\dfrac{u - u_C}{\bar{u}} = -\sqrt{\dfrac{c_f}{8}} \; f(\bar{y}_W)$	$0.15 < \bar{y}_W \le 1$

Kays and Crawford (1993)	$u^+ = y^+$	$0 \le y^+ < 5$
	$u^+ = 5\ln y^+ - 3.05$	$5 \le y^+ < 30$
Reichhard (1951)	$u^+ = 5.5 + 2.5\ln\left[y^+ \dfrac{1.5(1+\tilde{n})}{1+2\tilde{n}}\right]$	$y^+ \ge 30$
Mirushina and Ogisio (1970)	$u^+ = \dfrac{1}{6A^{2/3}}\left(A^{1/3} + \dfrac{1}{\mathrm{Re}_\tau}\right)\ln\left|\dfrac{\left(y^+ + 1/A^{1/3}\right)^3}{y^{+3} + 1/A}\right| +$	$0 \le y^+ \le y_1^+$
	$\quad + \dfrac{1}{\sqrt{3}A^{2/3}}\left(A^{1/3} - \dfrac{1}{\mathrm{Re}_\tau}\right) \times \tan^{-1}\left(\dfrac{2y^+ - 1/A^{1/3}}{\sqrt{3}A^{1/3}} + \dfrac{\pi}{6}\right)$	
	$u^+ = 2.5\ln y^+ + 5.5$	$y_1^+ < y^+ \le y_2^+$
	$u^+ = \dfrac{y^+ - y_2^+}{1 + 0.07\,\mathrm{Re}_\tau}\left(1 - \dfrac{y^+ + y_2^+}{2\,\mathrm{Re}_\tau}\right) + 2.5\ln y_2^+ + 5.5,$	$y_2^+ < y^+ \le \mathrm{Re}_\tau$
	$y_2^+ = 0.23\,\mathrm{Re}_\tau$	
	$A(y_1^+)^3 = 0.4 y_1^+\left(1 - \dfrac{y_1^+}{\mathrm{Re}_\tau}\right) - 1, \quad y_1^+ = 26.3$	

Appendix B: The Fully Developed Velocity Profile in an Axially Rotating Pipe

In the following appendix, the hydrodynamically fully developed velocity profile for a turbulent flow in an axially rotating pipe is discussed. The geometry under consideration is shown in Fig. B.1

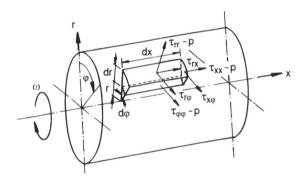

Fig. B.1: Geometry and coordinate system (Reich and Beer (1989))

For the flow in the axially rotating pipe, the velocity components u, v and w denote the flow in the x, r and φ direction.

For the following analysis, we restrict our considerations to an incompressible flow with constant fluid properties. The Reynolds averaged Navier-Stokes equations and the continuity equation, assuming rotational symmetry, are given for example by Rothe (1994)

$$\rho\left(\frac{\partial(uu)}{\partial x}+\frac{1}{r}\frac{\partial(ruv)}{\partial r}\right)=-\frac{\partial p}{\partial x}+\frac{\mu}{r}\frac{\partial}{\partial r}\left(r\frac{\partial u}{\partial r}\right)+\mu\frac{\partial^2 u}{\partial x^2}$$

$$-\frac{\rho}{r}\frac{\partial}{\partial r}\left(r\overline{u'v'}\right)-\rho\frac{\partial}{\partial x}\left(\overline{u'u'}\right)$$

(B.1)

$$\rho\left(\frac{\partial(uv)}{\partial x}+\frac{1}{r}\frac{\partial(rvv)}{\partial r}-\frac{w^2}{r}\right)=-\frac{\partial p}{\partial r}+\frac{\mu}{r}\frac{\partial}{\partial r}\left(r\frac{\partial v}{\partial r}\right)+\mu\frac{\partial^2 v}{\partial x^2}-\frac{v}{r^2} \qquad (B.2)$$

$$-\frac{\rho}{r}\frac{\partial}{\partial r}\left(r\overline{v'v'}\right)-\rho\frac{\partial}{\partial x}\left(\overline{u'v'}\right)-\rho\frac{\overline{w'w'}}{r}$$

$$\rho\left(\frac{\partial(uw)}{\partial x}+\frac{1}{r}\frac{\partial(rvw)}{\partial r}+\frac{vw}{r}\right)=\frac{\mu}{r}\frac{\partial}{\partial r}\left(r\frac{\partial w}{\partial r}\right)+\mu\frac{\partial^2 w}{\partial x^2}-\frac{w}{r^2} \qquad (B.3)$$

$$-\frac{\rho}{r}\frac{\partial}{\partial r}\left(r\overline{v'w'}\right)-\rho\frac{\partial}{\partial x}\left(\overline{u'w'}\right)-\rho\frac{\overline{v'w'}}{r}$$

$$\frac{\partial u}{\partial x}+\frac{1}{r}\frac{\partial}{\partial r}(rv)=0 \qquad (B.4)$$

For a hydrodynamically fully developed flow, the axial velocity component does not change with the axial position and, therefore, $\partial u/\partial x$ is zero. If we introduce this result into the continuity equation (B.4), we obtain for this case the result that the radial velocity component v is equal to zero and the continuity equation is then automatically satisfied. The fact that $v = 0$ for the fully developed flow is obvious, because a non-zero velocity component v would lead automatically to a change in the axial velocity component for different axial positions. Introducing this results into the above given equations result in

$$-\rho\frac{w^2}{r}=-\frac{\partial p}{\partial r}-\frac{\rho}{r}\frac{\partial}{\partial r}\left(r\overline{v'v'}\right)+\rho\frac{\overline{w'w'}}{r} \qquad (3.103)$$

$$0=\frac{1}{r^2}\frac{\partial}{\partial r}\left(\mu r^3\frac{\partial}{\partial r}\left(\frac{w}{r}\right)-\rho r^2\overline{v'w'}\right) \qquad (3.104)$$

$$0=\frac{\partial p}{\partial x}+\frac{1}{r}\frac{\partial}{\partial r}\left(\mu r\frac{\partial u}{\partial r}-\rho r\overline{u'v'}\right) \qquad (3.105)$$

with the boundary conditions

$$r=0:w=0,\ v=0,\ \frac{\partial u}{\partial r}=0 \qquad (3.106)$$

$$r=R:w=w_W,\ u=0,\ v=0$$

As mentioned in Chap. 3, experimental data show that the tangential velocity distribution is universal and can be approximated by Eq. (3.102).

$$\frac{w}{w_W} = \left(\frac{r}{R}\right)^2 \tag{3.102}$$

Figure 3.14 shows the tangential velocity distribution for different rotation rates N as well as for different Reynolds numbers. From Fig. 3.14, it is obvious that Eq. (3.102) is a very good approximation for the tangential velocity distribution. The axial velocity distribution $u(r)$ can be calculated from Eq. (3.105). In order to do this, the turbulent shear stress in this equation has to be related to the mean velocity gradients. This can be done by using a mixing length model according to Koosinlin et al. (1975). This results in the following expression for the turbulent shear stress

$$\rho \overline{u'v'} = \rho \, l^2 \left[\left(\frac{\partial u}{\partial r}\right)^2 + \left(r \frac{\partial}{\partial r}\left(\frac{w}{r}\right)\right)^2 \right]^{1/2} \frac{\partial u}{\partial r} = \varepsilon_m \rho \frac{\partial u}{\partial r} \tag{3.107}$$

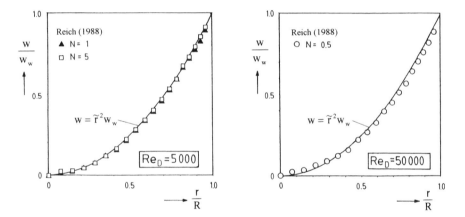

Fig. 3.14: Tangential velocity distribution (Reich and Beer (1989))

where the mixing length distribution l is given by (Reich and Beer (1989))

$$\frac{l}{l_0} = \left(1 - \frac{1}{6} \, \mathrm{Ri}\right)^2 \tag{3.108}$$

The mixing length distribution l_0 is the one for a non-rotating pipe (see Appendix A, Eq. (A.22)). The Richardson number in Eq. (3.108) describes the effect of pipe rotation on the turbulent motion and is defined as

$$Ri = \frac{2\dfrac{w}{r}\dfrac{\partial}{\partial r}(wr)}{\left(\dfrac{\partial u}{\partial r}\right)^2 + \left(r\dfrac{\partial}{\partial r}\left(\dfrac{w}{r}\right)\right)^2} \tag{3.109}$$

Without rotation $Ri = 0$, there exists a fully developed turbulent pipe flow. If $Ri > 0$, i.e. for an axially rotating pipe with a radially growing tangential velocity, the centrifugal forces suppress the turbulent fluctuations and the mixing length decreases. If we introduce the expression for the turbulent shear stress Eq. (3.107) and Eq. (3.102) into Eq. (3.105), we obtain the following nonlinear ordinary differential equation after integration

$$0 = \tilde{l}^2\left[\left(\frac{du^+}{d\tilde{r}}\right)^2 + \left(\tilde{r}\,N\frac{Re_D}{2\,Re_\tau}\right)^2\right]^{1/2}\left(\frac{du^+}{d\tilde{r}}\right) + \frac{1}{Re_\tau}\left(\frac{du^+}{d\tilde{r}}\right) + \tilde{r} \tag{B.5}$$

where the following dimensionless quantities have been used

$$\tilde{r} = \frac{r}{R}, \quad \tilde{l} = \frac{l}{R}, \quad u^+ = \frac{u}{u_\tau}, \quad u_\tau = \sqrt{|\tau_w|/\rho}, \tag{B.6}$$

$$Re_\tau = \frac{u_\tau R}{\nu}, \quad Re_D = \frac{\bar{u}\,D}{\nu}, \quad N = \frac{w_W}{\bar{u}}$$

In Eq. (B.5), the partial differentials have been replaced by ordinary differentials, because the axial velocity is only a function of the radial coordinate. Equation (B.5) is a strongly nonlinear ordinary differential equation and has to be solved with the boundary condition

$$u^+\left(\tilde{r} = 1\right) = 0 \tag{B.7}$$

Equation (B.5) together with Eq. (B.7) can now be solved in the following way: for a given value of the Reynolds number Re_D, a value of the shear stress Reynolds number Re_τ is assumed. After this, Eq. (B.5) can be integrated numerically so that the boundary condition according to Eq. (B.7) is fulfilled. After the integration, the continuity equation in integral form

$$\frac{Re_D}{4\,Re_\tau} = \int_0^1 u^+\tilde{r}\,d\tilde{r} \tag{B.8}$$

has to be satisfied. If Eq. (B.8) is not satisfied, a new value for Re_τ can be calculated from the equation. After some iteration, the desired velocity profile is obtained. The resulting axial velocity distribution is shown in Fig. 3.15.

As it can be seen from Fig. 3.15, the agreement between the experimental data and the numerical predictions is good.

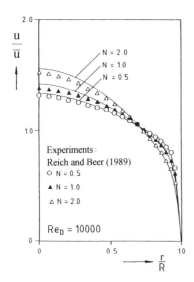
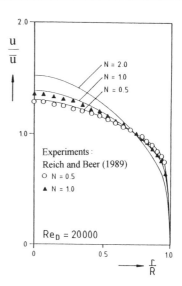

Fig. 3.15: Axial velocity distribution as a function of the rotation rate N (Reich and Beer (1989))

For turbulent flow in a non-rotating pipe, it is well known that the axial velocity profile can be described by an universal velocity profile, given by

$$\frac{u_C - u}{u_\tau} = f(\tilde{r})$$ (B.9)

In this equation, u_C denotes the axial velocity in the pipe center. The velocity law is valid over a large portion of the pipe radius and the function f is not dependent on the Reynolds number. If one examines Eq. (B.5) in more detail, one sees that only one quantity describes the influence of the rotation on the velocity field. This quantity is

$$Z = N \frac{Re_D}{2\,Re_\tau} = N / \sqrt{\frac{c_f}{8}}$$ (B.10)

From Eq. (B.10), it can be seen that the parameter Z involves the rotation rate N and the coefficient of friction loss c_f. The rotation rate N characterizes the effect of rotation on the axial mean velocity in the rotating pipe. The additional term $\sqrt{c_f / 8}$ takes into account the variation in the shape of the axial velocity profile at the pipe wall and, therefore, includes the effect of different pressure losses in the pipe section due to rotation.

Figure B.2 shows the functional relationship between Z and Re_D for various N. It can be seen that for a given value of N, the quantity Z increases with increasing values of the flow-rate Reynolds number. This is caused by the decreasing value of the friction factor with growing values of Re_D.

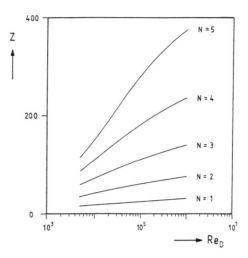

Fig. B.2: Rotation parameter Z as a function of the flow-rate Reynolds number with N as parameter (Weigand and Beer (1994))

If we assume in Eq. (B.5) to be far away from the wall so that $1/\mathrm{Re}_\tau\, du^+/d\tilde{r}$ is sufficiently small to be neglected in comparison with all other terms in this equation, we can hope to find an universal velocity law for the core region if the mixing length is only a function of $\tilde{r}$ and Z. From the above assumption, it is clear that the following analysis is only valid if N does not approach infinity. In this latter case, the whole pipe cross section is influenced by the viscous forces and the axial velocity distribution tends to the one for laminar pipe flow (Hagen-Poiseuille flow)

$$\frac{u}{\bar{u}} = 2\left(1 - \tilde{r}^2\right) \tag{B.11}$$

This behavior is caused by the vanishing turbulent shear stress with increasing rotation rate N. If we now exclude extremely large values of N and introduce the dimensionless quantities defined in Eq. (B.6), we obtain from Eq. (3.109) for the Richardson number,

$$\mathrm{Ri} = \frac{6\tilde{r}^2 Z^2}{\left(\dfrac{\partial u^+}{\partial \tilde{r}}\right)^2 + \left(\tilde{r}Z\right)^2} \tag{B.12}$$

and the mixing length distribution is given by Eq. (3.108).

Because the mixing length only depends on the radial coordinate and the Richardson number, it follows from Eq. (B.5) that the velocity profile in the core region must have an universal character. Figure B.3 shows the distribution of the axial velocity for various values of Z. It can be seen that the velocity profiles are

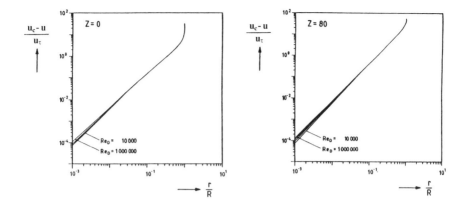

Fig. B.3: Axial velocity distribution for various values of the rotation rate Z and different Re_D (Weigand and Beer (1994))

universal. There is only a difference in the profiles for different Reynolds numbers near the wall ($\tilde{r} \approx 1$), where the viscous forces play an important role.

In order to derive an approximation for the axial velocity distribution, we follow the work of Reichardt (1951) for the non-rotating pipe and subdivide the flow area $(0 \leq \tilde{r} \leq 1)$ into two parts: a near wall region and a part which contains the rest of the flow area. Reichardt (1951) developed, for the turbulent flow in a non-rotating pipe, an approximation for the velocity distribution which is valid for the whole flow area as well. This is also the aim for the case of turbulent flow in an axially rotating pipe. For more details on this subject, the reader is referred to Weigand and Beer (1994).

The Velocity Distribution in the Core Region:

If one plots the velocity distribution of Fig. B.3 in a diagram containing logarithmic axis, one obtains for the velocity distribution

$$\frac{u_C - u}{u_\tau} = A\tilde{r}^B \tag{B.13}$$

where A and B are depending on Z

$$A(Z) = \sqrt{0.06052\,Z^2 + 5.4\,Z + 25.705}, \tag{B.14}$$
$$B(Z) = 1.55 + 0.338\left[1 - \exp\left(-0.0553\,Z\right)\right]$$

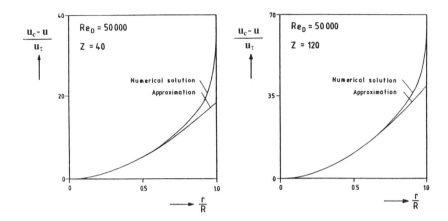

Fig. B.4: Comparison between numerical solution and the approximation for the axial velocity in the core region (Weigand and Beer (1994))

For Z = 0 (non-rotating pipe), the velocity distribution given by the Eqs. (B.13-B.14) is identical to the one reported by Darcy (1858). Figure B.4 shows a comparison between the numerically predicted axial velocity distribution and the approximation given by Eqs. (B.13-B.14). It can be seen that Eqs. (B.13-B.14) approximate the velocity profile very well for $\tilde{r} \le 0.7$. For larger values of $\tilde{r}$, larger deviations can be noticed.

The Velocity Distribution in the Near Wall Region:

It is well known that the velocity distribution near the wall in a non-rotating pipe has a logarithmic distribution given by

$$u^+ = \frac{1}{\kappa} \ln y^+ + c \tag{A.36}$$

where κ and c are constants and y^+ is the dimensionless distance from the wall, given by Eq. (A.23). In a rotating pipe, the turbulent fluctuations are suppressed by the centrifugal forces. Furthermore, the area, where the velocity profile is still logarithmic in shape, decreases with increasing rotation rates. This fact is shown in Fig. B.5 and has been observed in DNS calculations by Eggels and Nieuwstadt (1993) and by Orlandi and Fatica (1995). However, Weigand and Beer (1994) obtained an approximation formula for the velocity distribution in the near wall region, which includes the effect of relaminarization due to system rotation.

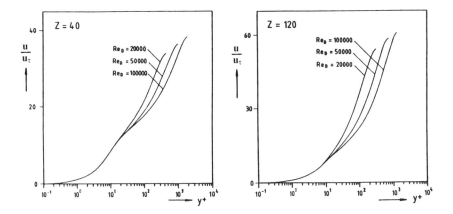

Fig. B.5: Axial velocity distribution u/u_τ as a function of the wall coordinate y^+ for two different values of Z (Weigand and Beer (1994))

This equation is given by (for more details the reader is referred to Weigand and Beer (1994))

$$u^+ = \frac{1}{\kappa_1}\ln\left(1+\kappa_1 y^+\right)+7.8\left[\frac{\mathrm{Re}_\tau}{\left(\mathrm{Re}_\tau\right)_0}\left(1-\exp\left(\frac{y^+}{y_1^+}\right)-\left(\frac{y^+}{y_1^+}\right)\exp\left(-by^+\right)\right)\right] \quad (\text{B.15})$$

with the quantities

$$y_1^+ = y_{10}\frac{\mathrm{Re}_\tau}{\left(\mathrm{Re}_\tau\right)_0}, \quad y_{10}=11$$

$$\kappa_1 = \kappa\left(\frac{\mathrm{Re}_\tau}{\left(\mathrm{Re}_\tau\right)_0}\right)/\left[\left(\frac{Z}{\mathrm{Re}_\tau}\right)^2+1\right]^{3.5}, \quad \kappa=0.4 \quad (\text{B.16})$$

$$b = \frac{1}{3}\sqrt{\frac{\kappa_1}{\kappa}}$$

In the equations above $\left(\mathrm{Re}_\tau\right)_0$ denotes the shear-stress Reynolds number for the non-rotating pipe. As it can be seen from Eq. (B.15), the velocity distribution in the near wall region is influenced by a term containing the influence of the system rotation on the wall shear stress. Figure B.6 shows the good agreement between the approximation given above and the numerical solution of the problem.

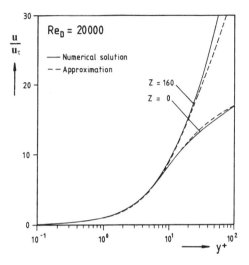

Fig. B.6: Comparison between numerical solution and the approximation for u/u_τ in the near wall region (Weigand and Beer (1994))

An Approximation Formula for the Whole Region of the Pipe:

Strictly speaking, the two equations given above for the axial velocity distribution are only valid in the near wall region and in the core region. However, the two solutions can be combined in one single relation, as shown by Weigand and Beer (1994). They obtained

$$u^+ = \left\{ \frac{1}{\kappa_1} \ln\left(1 + \kappa_1 y^+\right) + 7.8 \left[\frac{\mathrm{Re}_\tau}{(\mathrm{Re}_\tau)_0} \left(1 - \exp\left(\frac{y^+}{y_1^+}\right) - \left(\frac{y^+}{y_1^+}\right) \exp\left(-by^+\right) \right) \right] \right\} \quad (B.17)$$

$$\left\{ \exp\left(a_1\left(1 - \tilde{r}\right)\right) \right\} + \left\{ \frac{u_C}{u_\tau} - A\tilde{r}^B \right\} \left\{ 1 - \exp\left(a_1\left(1 - \tilde{r}\right)\right) \right\}$$

with $a_1 = 5\left(\mathrm{Re}_\tau\right)_0 / \mathrm{Re}_\tau$. From the continuity equation in integral form, one can finally obtain a functional dependence between the shear-stress velocity and the maximum velocity in the pipe center

$$\frac{u_C}{u_\tau} = \frac{\mathrm{Re}_D}{2\,\mathrm{Re}_\tau} + \frac{2A}{2+B} \quad (B.18)$$

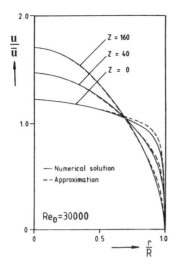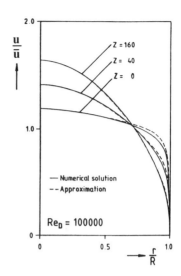

Fig. B.7: Influence of a variation of the rotation parameter Z on the shape of the axial velocity profile (Weigand and Beer (1994))

Fig. B.7 elucidates the influence of the rotation parameter Z on the axial velocity distribution for two different flow-rate Reynolds numbers. It can be seen that increasing Z tends to relaminarize the flow. The axial velocity profiles tend to approach the parabolic distribution of the Hagen-Poiseuille flow for growing values of Z. Furthermore, it can be seen that the above obtained equation for the axial velocity approximates very well the numerical calculation.

If the approximation is used in Fig.3.15 instead of the numerical solution, no noticeable difference can be seen.

Finally, it should be noted here that the effect of pipe rotation is very much different if a laminar flow enters the axially rotating pipe. In contrast to the previous discussions, the rotation causes for a laminar inlet flow a destabilization of the flow, even so the flow rate Reynolds number is much smaller than 2000. Mackrodt (1971) found that a laminar pipe flow is only stable against circumferential disturbances if the Reynolds numbers are below the limits $Re_\varphi = 53.92$ and $Re_D = 165.76$. The reader is referred to the papers of Murakami and Kikuyama (1980), Kikuyama et al. (1983), Reich et al. (1989) and Weigand and Beer (1992b) for a more detailed discussion of this subject.

Appendix C: A Numerical Solution Method for Eigenvalue Problems

During the analytical solution process for the thermal entrance problems in pipe and channel flows, a linear partial differential equation of the kind

$$\tilde{u}(\tilde{n})\frac{\partial \Theta}{\partial \tilde{x}} = \frac{1}{\tilde{r}^F}\frac{\partial}{\partial \tilde{n}}\left[\tilde{r}^F a_2(\tilde{n})\frac{\partial \Theta}{\partial \tilde{n}}\right] \tag{3.19}$$

has to be solved. If we consider a step change in the wall temperature, the boundary conditions for this partial differential equation are given by

$$\tilde{x} = 0 : \Theta(0,\tilde{n}) = 1 \tag{3.17}$$

$$\tilde{n} = 0 : \frac{\partial \Theta}{\partial \tilde{n}}\bigg|_{\tilde{n}=0} = 0$$

$$\tilde{n} = 1 : \Theta(\tilde{x},1) = 0$$

The solution of the above given problem has been derived in Chap. 3 and is given by

$$\Theta = \sum_{j=0}^{\infty} A_j \Phi_j(\tilde{n}) \exp\left(-\lambda_j^2 \tilde{x}\right) \tag{3.55}$$

In the following appendix, one possible numerical solution procedure for the eigenvalue problems discussed in Chap. 3 and Chap. 4 will be presented. As it has been demonstrated, the solution of the energy equation can be obtained for these problems as a sum of eigenfunctions. In order to determine the temperature distribution, the eigenfunctions and the related eigenvalues have to be predicted. For the above given problem (Eqs. (3.17, 3.19)) the eigenfunctions are the solutions of the following ordinary differential equation

$$\tilde{r}^F \tilde{u}(\tilde{n})\lambda_j^2 \Phi_j(\tilde{n}) + \left[\tilde{r}^F a_2(\tilde{n})\Phi_j'(\tilde{n})\right]' = 0 \tag{3.25}$$

with the boundary conditions

$$\Phi_j'(0) = 0 \tag{3.26}$$

$$\Phi_j(1) = 0$$

The flow index F in Eq. (3.25) has to be set to one for the heat transfer in pipe flows and to zero for the heat transfer in a parallel plate channel.

Eigenvalue problems, like the one given by the Eqs. (3.25-3.26), can normally only be solved numerically, because the coefficients in the differential equation are complicated functions of the coordinate $\tilde{n}$. This is the case for the velocity distribution $\tilde{u}(\tilde{n})$ and also for the function

$$a_2(\tilde{n}) = 1 + \frac{\text{Pr}}{\text{Pr}_t}\tilde{\varepsilon}_m \qquad (C.1)$$

One efficient method to solve these eigenvalue problems is by using a Runge-Kutta method. This approach will be explained now in detail:

In order to solve the Eqs. (3.25-3.26) by a Runge-Kutta method, we transform the second-order ordinary differential equation into a system of two first-order ordinary differential equation. Introducing the new function

$$\chi_j = \tilde{r}^F a_2(\tilde{n})\Phi'_j \qquad (C.2)$$

into Eq. (3.25) results in the following system of ordinary differential equations

$$\begin{pmatrix} \Phi'_j \\ \chi'_j \end{pmatrix} = \begin{pmatrix} \chi_j/(\tilde{r}^F a_2(\tilde{n})) \\ -\tilde{r}^F \tilde{u} \lambda_j^2 \Phi_j \end{pmatrix} = \begin{pmatrix} 0 & (\tilde{r}^F a_2)^{-1} \\ -\tilde{r}^F \tilde{u} \lambda_j^2 & 0 \end{pmatrix}\begin{pmatrix} \Phi_j \\ \chi_j \end{pmatrix} \qquad (C.3)$$

In order to solve the system of equations given by Eq. (C.3) with a Runge-Kutta method, the problem has to be considered as an initial value problem. Therefore, we provide an additional normalizing condition for the eigenfunctions according to

$$\Phi_j(0) = 1 \qquad (C.4)$$

Using an arbitrary normalizing condition does not change the solution of the problem, because in Eq. (3.55) all the individual eigenfunctions are multiplied by constants. This means that the value prescribed by Eq. (C.4) will only influence the value of the constants A_j in Eq. (3.55).

The algorithm for the Runge-Kutta method for solving a system of ordinary differential equations is relatively simple (see for example Törnig (1979)). The system of differential equations, given by Eq. (C.3), is now solved together with the two initial conditions

$$\Phi_j(0) = 1 \qquad (C.5)$$
$$\chi_j(0) = 0$$

This will be done by using a guessed value of the eigenvalue λ_a^2. Examining the boundary condition for the eigenfunction for $\tilde{n} = 1$ ($\Phi_j(1) = 0$) shows, if the guessed value λ_a^2 is an eigenvalue of the problem under consideration. If λ_a^2 is not an eigenvalue, the calculation procedure will be continued with a different guessed value $\lambda_b^2 > \lambda_a^2$. This procedure, which is known in literature as a shooting method, will be continued up to the point when a change in sign appears for $\Phi_j(1)$. If this is the case, an eigenvalue has been found in the interval $\left[\lambda_a^2, \lambda_b^2 \right]$. This eigenvalue can then be predicted to whatever accuracy is desired, by successively halving the interval. This means that the eigenvalue can finally be predicted with the accuracy

$$\Phi_j(1) \le \varepsilon \tag{C.6}$$

where it is normally accurate enough to set $\varepsilon = 10^{-6}$. The here discussed procedure is illustrated in Fig. C.1.

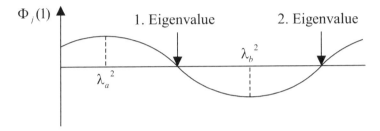

Fig. C.1: Numerical prediction of the eigenvalues

The number of zero-points of the eigenfunctions Φ_j increases linearly with growing values of j. This means that higher eigenfunctions strongly oscillate and that it is difficult to capture exactly the shape of the eigenfunctions for larger j by a numerical method. In Chap. 4, we have investigated an analytical methods for predicting such eigenfunctions for large values of j. However, if the number of grid points used to resolve the interval $[0,1]$ in the numerical prediction is large enough, also higher eigenfunctions and eigenvalues can be predicted accurately. For this normally $1000 - 2000$ grid points have shown to be sufficient.

C.1 Numerical Tools

On the internet web page: www.uni-stuttgart.de/itlr/ the reader will find a page entitled *Analytical Methods*. The reader can get the login and the password for this page upon request from the author (bw@itlr.uni-stuttgart.de). Here are several useful programs (executables and source codes) for the prediction and animation of thermal entrance heat transfer problems.

The programs for predicting thermal entrance problems in a parallel plate channel will be explained here in a little more detail, because for these programs also the source code is provided on the web page stated above.

Vel2dch

This program (Velocity-Profile-2D-Channel), which is written in FORTRAN, predicts the hydrodynamically fully developed velocity profile for a laminar or turbulent channel flow. The underlying equations, which are solved, are discussed in detail in Appendix A. At the start of the program the reader is asked to provide a Reynolds number for the channel flow. If a value for the Reynolds number, based on the hydraulic diameter is provided, which is smaller than 2000, the program automatically calculates the velocity profile for a laminar channel flow (see Eq. (3.2)).

In addition to the solution of the hydrodynamically fully developed velocity profile, the program also predicts the function given by Eq. (C.1) and the fully developed temperature distribution in case of a constant wall heat flux. For the turbulent heat diffusivity, the turbulent Prandtl number model by Weigand et al. (1997a) is used. For laminar flow, the function $a_2(\tilde{n}) = 1$, and the fully developed temperature distribution is given by Eq. (5.76) (for $\text{Pe}_L \rightarrow \infty$). This program serves as an input generator to the program *Temp2dch* (Temperature-Distribution-2D-Channel), which solves the energy equation for the thermal entrance problem in a parallel plate channel.

Temp2dch

This program, which is also written in FORTRAN, predicts the temperature distribution and the Nusselt number for a hydrodynamically fully developed flow. The underlying equations are discussed in detail in Chap. 3. Solutions of the energy equation are provided for the case of constant wall temperature and for the case of constant wall heat flux. The program uses the velocity distribution, the function $a_2(\tilde{y})$ and the fully developed temperature profile (which is only needed in case of a constant wall heat flux boundary condition) from the program *Vel2dch*. The calculation of the eigenvalues is done by using a Runge-Kutta method and follows the approach outlined in the first part of this chapter.

Visualization of Results

After executing the above described programs, several data files are generated, which can be visualized by freeware programs as explained on the previous mentioned web page. The output files contain all main interesting calculated quantities like temperature profiles (line plots and the development of the profiles as a function of axial coordinate), Nusselt number distribution, etc..

All the programs provided on the internet page have been rewritten from older FORTRAN programs in order to be more user friendly. The author has used similar programs for a long time and a lot of comparisons have been made over time between predicted eigenvalues, constants, Nusselt number distributions and literature data. This includes comparisons for pipe flow with Shah (1975), Shah and London (1978), Papoutsakis et al. (1980a) (with axial heat conduction), comparisons for parallel plate channels with Brown (1960), Deavours (1974), Shah and London (1978) and for the case of circular annuli with Lundberg et al. (1963) and with Hsu (1970) (with axial heat conduction).

Table C.1 shows a comparison of present predictions with values reported by Shah (1975) for a laminar flow in a parallel plate channel with constant wall temperature. It can be seen that the present calculations are in excellent agreement with the values of Shah (1975). This is interesting, because Shah (1975) obtained for $x/(D\,\mathrm{Pe}_D) \leq 10^{-4}$ the values for the Nusselt number from an extended Leveque solution[1].

Table C.1: Comparison of local Nusselt numbers for the thermal entrance region in a parallel plate channel with laminar internal flow and constant wall temperature

$x/(D\,\mathrm{Pe}_D)$	Nu_D (Shah (1975))	Nu_D (own calculation)
0.000001	122.943	122.7618
0.000003	85.187	85.2384
0.000005	71.830	71.8536
0.000006	67.589	67.6069
0.000007	64.200	64.2141
0.000008	61.403	61.4139
0.000009	59.037	59.0463
0.00001	56.999	57.0064
0.00003	39.539	39.5402
0.00005	33.379	33.3791
0.0001	26.560	26.5600
0.0005	15.830	15.8300
0.001	12.822	12.8217
0.005	8.5166	8.5166
0.01	7.7405	7.7405
0.05	7.5407	7.5407
0.1	7.5407	7.5407

In addition, this sort of programs have been compared against numerical predictions for pipe flows (for example Hennecke (1968)) and for circular annuli (Fuller and Samuels (1971)). Some of the above mentioned comparisons have already been shown in Chap. 3 and Chap. 4. The reader will find it relatively straight forward to extend the programs provided on the above given web page to

---

[1] The series solution for the temperature field and for the Nusselt number, which have been obtained in Chap. 3, are converging very slowly for $\tilde{x} \to 0$. Leveque (1928) developed an approximation for the temperature field for very small values of $\tilde{x}$. He approximated the velocity profile, within the very thin thermal boundary layer by $\tilde{u} = 1 - \tilde{n}$. By doing so, the energy equation has a similarity solution (see Chap. 6). This solution can be used as long as the thermal boundary layer thickness is very thin ($x/(D\,\mathrm{Pe}_D) \leq 10^{-3}$). The reader is also referred to Worsoe-Schmitt (1967) and to Kader (1971), where also solutions of this kind are reported for the extended Graetz problem (with axial heat conduction).

the case of pipe flow and flow in circular annuli. Also the extension to solve eigenvalue problems for situations, where the axial heat conduction cannot be ignored, is relatively easy.

One Example

On the web-page the reader will find also some example calculations and a documentation in order to show how to use the programs. In addition also some comparisons between the programs and literature data are provided. One simple example, which could be used as a start, in order to get used to the programs, could be to compute the laminar heat transfer for a slug-flow velocity profile. Here one can develop very easily a complete analytical solution. Let us assume that

$$\tilde{u}(\tilde{y}) = 1 \qquad (C.7)$$

This could be the case for a laminar flow of a liquid metal with a very low Prandtl number. Inserting Eq. (C.7) into the energy equation (3.15) results in

$$\frac{\partial \Theta}{\partial \tilde{x}} = \frac{\partial^2 \Theta}{\partial \tilde{y}^2} \qquad (C.8)$$

Let us now consider constant wall temperature boundary conditions according to Eq. (3.17). Then we obtain for the planar channel

$$\tilde{x} = 0 : \Theta(0, \tilde{y}) = 1 \qquad (3.17)$$

$$\tilde{y} = 0 : \frac{\partial \Theta}{\partial y}\bigg|_{\tilde{y}=0} = 0$$

$$\tilde{y} = 1 : \Theta(\tilde{x}, 1) = 0$$

The solution of Eq. (C.8) together with the boundary conditions given by Eq. (3.17) can be easily obtained by the method of separation of constants and one gets

$$\Theta = \sum_{j=0}^{\infty} A_j \cos(\lambda_j \tilde{y}) \exp(-\lambda_j^2 \tilde{x}) \qquad (C.9)$$

with the quantities

$$A_j = \frac{2 \sin(\lambda_j)}{\lambda_j}, \ \lambda_j = \frac{(2j+1)\pi}{2}, \ j = 0,1,2,... \qquad (C.10)$$

The uniform velocity profile given by Eq. (C.7) can easily been modified in the input file for the program *Temp2dch* by setting all values for the velocity profile to one.

A comparison between the analytical predicted eigenvalues and the numerical calculated values is reported in Table C2. It can be seen that there is an excellent agreement between the numerically predicted values and the exact solution.

Table C.2: Comparison between analytically predicted eigenvalues and calculated eigenvalues by using the program *Temp2dch*

Nr.	$(\lambda_j^2)_{analytical}$	$(\lambda_j^2)_{numerical}$	Rel. Error
0	2.4674012	2.4674009	$< 10^{-7}$
1	22.206611	22.206608	$< 10^{-7}$
2	61.685031	61.685022	$< 10^{-7}$
3	120.90266	120.90264	$< 10^{-7}$
4	199.85950	199.85947	$< 10^{-7}$
5	298.55555	298.55551	$< 10^{-7}$
6	416.99081	416.99075	$< 10^{-7}$
7	555.16528	555.16520	$< 10^{-7}$
8	890.73185	890.73172	$< 10^{-7}$

The same sort of accuracy can also be achieved for the constants A_j. A comparison between numerically predicted values and the exact solution is given in Table C.3

Table C.3: Comparison between analytically predicted constants A_j and numerically calculated values by using the program *Temp2dch*

Nr.	$(A_j)_{analytical}$	$(A_j)_{numerical}$	Rel. Error
0	1.2732395	1.2732395	$< 5 \; 10^{-8}$
1	-0.42441317	-0.42441313	$< 1 \; 10^{-7}$
2	0.25464790	0.25464783	$< 3 \; 10^{-7}$
3	-0.18189136	-0.18189125	$< 7 \; 10^{-7}$
4	0.14147106	0.14147091	$< 2 \; 10^{-6}$
5	-0.11574905	-0.11574887	$< 2 \; 10^{-6}$
6	0.097941501	0.097941291	$< 3 \; 10^{-6}$
7	-0.084882634	-0.084882391	$< 3 \; 10^{-6}$
8	0.074896442	0.074896166	$< 4 \; 10^{-6}$

More examples can be found on the previous mentioned web-page.

Please note that the provided programs on the web-page are for educational use only. They should not be used for any commercial applications. Furthermore, the author does not take any warranty for any results produced by the programs.

Appendix D: Detailed Derivation of Certain Properties of the Method for Solving the Extended Graetz Problems

In the following appendix, certain properties of the method explained in Chap. 5 are derived in detail. Where necessary, the derivation is shown for the two different boundary conditions considered in Chap. 5: constant wall temperature and constant wall heat flux.

D.1 Symmetry of the Matrix Operator $\underset{\sim}{L}$

We start our considerations by showing that the matrix operator $\underset{\sim}{L}$ is a symmetric operator in the Hilbert space $H = H_1 \oplus H_2$, where H_1 is the space containing all functions $f(\tilde{n})$ in $[0,1]$ for which the integral

$$\int_0^1 f^2(\tilde{n}) \tilde{r}^F a_1(\tilde{n}) d\tilde{n} < \infty \tag{D.1}$$

holds. H_2 is the space of all functions $b(\tilde{n})$ in $[0,1]$ for which Eq. (D.2) holds

$$\int_0^1 \frac{b^2(\tilde{n})}{\tilde{r}^F a_2(\tilde{n})} d\tilde{n} < \infty \tag{D.2}$$

Now it can be shown, that the matrix operator $\underset{\sim}{L}$, given by Eq. (5.11), is symmetric with respect to the inner product defined by Eq. (5.14). We assume that the two vectors $\vec{\Phi}, \vec{\Upsilon}$ belong to $D(\underset{\sim}{L})$, which is defined for a given wall temperature by Eq. (5.15) and for a given wall heat flux by Eq. (5.59). This means that we have to show, that

$$\left\langle \vec{\Phi}, \underset{\sim}{L}\vec{\Upsilon} \right\rangle = \left\langle \underset{\sim}{L}\vec{\Phi}, \vec{\Upsilon} \right\rangle \tag{5.16}$$

Equation (5.16) can be proven by inserting the expressions $\underset{\sim}{L}\vec{\Phi}$ and $\underset{\sim}{L}\vec{\Upsilon}$ into the definition of the inner product according to Eq. (5.14). This results in

$$\left\langle \vec{\Phi}, \underset{\sim}{L}\vec{\Upsilon} \right\rangle = \int\limits_0^1 \left[\frac{\tilde{u}\,\tilde{r}^F}{Pe_L}\Phi_1(\tilde{n})\,\Upsilon_1(\tilde{n}) - \Phi_1(\tilde{n})\Upsilon_2'(\tilde{n}) + \Phi_2(\tilde{n})\,\Upsilon_1'(\tilde{n}) \right] d\tilde{n} \qquad (5.17)$$

$$\left\langle \underset{\sim}{L}\vec{\Phi}, \vec{\Upsilon} \right\rangle = \int\limits_0^1 \left[\frac{\tilde{u}\,\tilde{r}^F}{Pe_L}\Phi_1(\tilde{n})\,\Upsilon_1(\tilde{n}) - \Phi_2'(\tilde{n})\Upsilon_1(\tilde{n}) + \Phi_1'(\tilde{n})\,\Upsilon_2(\tilde{n}) \right] d\tilde{n} \qquad (5.18)$$

Subtracting Eq. (5.18) from Eq. (5.17) results after integration in

$$\left\langle \vec{\Phi}, \underset{\sim}{L}\vec{\Upsilon} \right\rangle - \left\langle \underset{\sim}{L}\vec{\Phi}, \vec{\Upsilon} \right\rangle = \Phi_2(1)\,\Upsilon_1(1) - \Phi_2(0)\,\Upsilon_1(0) \qquad (5.19)$$
$$- \Phi_1(1)\,\Upsilon_2(1) + \Phi_1(0)\,\Upsilon_2(0)$$

The resulting expression on the right hand side of Eq. (5.19) is zero for the case of a given wall temperature because $\Phi_1(1) = \Upsilon_1(1) = 0$, as well as for the case of a given wall heat flux boundary condition, because $\Phi_2(1) = \Upsilon_2(1) = 0$. For both boundary conditions $\Phi_2(0) = \Upsilon_2(0) = 0$ because of the assumed symmetry of the eigenfunctions.

D.2 The Eigenfunctions Constitute a Set of Orthogonal Functions

Next, we show that the eigenfunctions given in Chap. 5 constitute a set of orthogonal functions. Consider the two eigenfunctions $\vec{\Phi}_j$ and $\vec{\Phi}_k$ together with the related eigenvalues λ_j and λ_k. From Eq. (5.20) one obtains:

$$\underset{\sim}{L}\vec{\Phi}_j = \lambda_j \vec{\Phi}_j \qquad (D.3)$$

$$\underset{\sim}{L}\vec{\Phi}_k = \lambda_k \vec{\Phi}_k$$

with the boundary conditions

Constant Wall Temperature

$$\Phi_{j1}'(0) = \Phi_{k1}'(0) = 0, \qquad \Phi_{j1}(1) = \Phi_{k1}(1) = 0 \qquad (D.4)$$

Constant Wall Heat Flux

$$\Phi_{j1}'(0) = \Phi_{k1}'(0) = 0, \qquad \Phi_{j1}'(1) = \Phi_{k1}'(1) = 0 \qquad (D.5)$$

By taking the inner product (defined by Eq. (5.14)) of Eq. (D.3), one obtains (note, that from Eq. (5.14) follows that $\left\langle \vec{\Phi}_j, \vec{\Phi}_k \right\rangle = \left\langle \vec{\Phi}_k, \vec{\Phi}_j \right\rangle$)

$$\left\langle \underset{\sim}{L}\vec{\Phi}_j,\vec{\Phi}_k \right\rangle - \left\langle \vec{\Phi}_j, \underset{\sim}{L}\vec{\Phi}_k \right\rangle = \left(\lambda_j - \lambda_k \right)\left\langle \vec{\Phi}_j,\vec{\Phi}_k \right\rangle \tag{D.6}$$

In the last paragraph, it has been shown that the operator $\underset{\sim}{L}$ is a symmetric operator in the Hilpert space. This means that Eq. (5.16) is valid

$$\left\langle \vec{\Phi}, \underset{\sim}{L}\vec{\Upsilon} \right\rangle = \left\langle \underset{\sim}{L}\vec{\Phi},\vec{\Upsilon} \right\rangle \tag{5.16}$$

where $\vec{\Phi}$ and $\vec{\Upsilon}$ can be replaced by $\vec{\Phi}_j$ and $\vec{\Phi}_k$. Then it follows immediately from Eq. (D.6) that

$$0 = \left(\lambda_j - \lambda_k \right)\left\langle \vec{\Phi}_j,\vec{\Phi}_k \right\rangle \tag{D.7}$$

From this equation, one can conclude that

$$\left\langle \vec{\Phi}_j,\vec{\Phi}_k \right\rangle = \begin{cases} 0 & \text{for } \lambda_j \neq \lambda_k \\ \left\| \vec{\Phi}_j \right\|^2 & \text{for } \lambda_j = \lambda_k \end{cases} \tag{D.8}$$

Eq. (D.8) shows that the eigenfunctions constitute a set of orthogonal functions for the inner product defined according to Eq. (5.14).

D.3 A detailed Derivation of Eq. (5.31) and Eq. (5.61)

Because of the non-homogeneous boundary conditions, the vector $\vec{S}$ does not belong to the domain $D(\underset{\sim}{L})$ given by Eq. (5.15) (for constant wall temperature boundary conditions) or by Eq. (5.59) (for constant wall heat flux boundary conditions). However, it can be shown that Eq. (5.31) and Eq. (5.61) hold.

$$\left\langle \underset{\sim}{L}\vec{S},\vec{\Phi}_j \right\rangle = \left\langle \vec{S}, \underset{\sim}{L}\vec{\Phi}_j \right\rangle + \Phi_{j2}(1)g(\tilde{x}) \tag{5.31}$$

$$\left\langle \underset{\sim}{L}\vec{S},\vec{\Phi}_j \right\rangle = \left\langle \vec{S}, \underset{\sim}{L}\vec{\Phi}_j \right\rangle - \Phi_{j1}(1)E\left(\tilde{x},1 \right) \tag{5.61}$$

These two equations can be derived in the following way: from the definition of the inner product by Eq. (5.14) one obtains

$$\left\langle \underset{\sim}{L}\vec{S},\vec{\Phi}_j \right\rangle = \int_0^1 \left[\frac{\tilde{u}\,\tilde{r}^F}{Pe_L}\Phi_{j1}(\tilde{n})S_1(\tilde{n}) - \Phi_{j1}(\tilde{n})S_2'(\tilde{n}) + \Phi_{j2}(\tilde{n})S_1'(\tilde{n}) \right]d\tilde{n} \tag{D.9}$$

$$\left\langle \vec{S}, \underset{\sim}{L}\vec{\Phi}_j \right\rangle = \int_0^1 \left[\frac{\tilde{u}\,\tilde{r}^F}{Pe_L}\Phi_{j1}(\tilde{n})S_1(\tilde{n}) - \Phi_{j2}'(\tilde{n})S_1(\tilde{n}) + \Phi_{j1}'(\tilde{n})S_2(\tilde{n}) \right]d\tilde{n} \tag{D.10}$$

where the prime denotes the derivative with respect to $\tilde{n}$. From these two equations one obtains

$$\left\langle \underset{\sim}{L}\vec{S},\vec{\Phi}_j \right\rangle = \left\langle \vec{S}, \underset{\sim}{L}\vec{\Phi}_j \right\rangle + \Phi_{j2}(1)S_1(1) - \Phi_{j2}(0)S_1(0) + \tag{D.11}$$
$$-\Phi_{j1}(1)S_2(1) + \Phi_{j1}(0)S_2(0)$$

where the solution vector $\vec{S}$ was defined in Eq. (5.11) by

$$\vec{S} = \begin{bmatrix} \Theta(\tilde{x},\tilde{n}) \\ E(\tilde{x},\tilde{n}) \end{bmatrix} \tag{D.12}$$

Therefore, Eq. (D.11) can also be written as

$$\left\langle \underset{\sim}{L}\vec{S},\vec{\Phi}_j \right\rangle = \left\langle \vec{S}, \underset{\sim}{L}\vec{\Phi}_j \right\rangle + \Phi_{j2}(1)\Theta(\tilde{x},1) - \Phi_{j2}(0)\Theta(\tilde{x},0) \tag{D.13}$$
$$-\Phi_{j1}(1)E(\tilde{x},1) + \Phi_{j1}(0)E(\tilde{x},0)$$

For the case of a constant wall temperature, the boundary conditions are given by

$$\Phi'_{j1}(0) = 0, \quad \Phi_{j1}(1) = 0, \tag{D.14}$$
$$\left.\frac{\partial \Theta}{\partial \tilde{n}}\right|_{\tilde{n}=0} = 0, \quad \Theta(\tilde{x},1) = \text{given}$$

In addition, it follows from Eq. (5.22) that $\Phi_{j2}(0) = 0$ and from the definition of the function E by Eq. (5.9) it follows that $E(\tilde{x},0) = 0$. Therefore, Eq. (D.13) simplifies for the case of a given wall temperature to

$$\left\langle \underset{\sim}{L}\vec{S},\vec{\Phi}_j \right\rangle = \left\langle \vec{S}, \underset{\sim}{L}\vec{\Phi}_j \right\rangle + \Phi_{j2}(1)\Theta(\tilde{x},1) \tag{D.15}$$

If now $\Theta(\tilde{x},1)$ is replaced by the boundary condition according to Eq. (5.7), Eq. (5.31) is obtained.

For the case of a given wall heat flux, the boundary conditions are given by

$$\Phi'_{j1}(0) = 0, \quad \Phi'_{j1}(1) = 0, \tag{D.16}$$
$$\left.\frac{\partial \Theta}{\partial \tilde{n}}\right|_{\tilde{n}=0} = 0, \quad \left.\frac{\partial \Theta}{\partial \tilde{n}}\right|_{\tilde{n}=1} = given$$

Remembering that $\Phi_{j2}(0) = 0$ and $E(\tilde{x},0) = 0$, one obtains from Eq. (D.13), by considering that $\Phi_{j2}(1) = 0$ for the case of a prescribed wall heat flux, Eq. (5.61)

$$\left\langle \underset{\sim}{L}\vec{S},\vec{\Phi}_j \right\rangle = \left\langle \vec{S}, \underset{\sim}{L}\vec{\Phi}_j \right\rangle - \Phi_{j1}(1)E(\tilde{x},1) \tag{5.61}$$

D.4 Simplification of the Expression for the Temperature Distribution (for Constant Wall Temperature)

In Eq. (5.42) we used the result that

$$-\sum_{j=0}^{\infty} \frac{\Phi_{j2}^-(1)\Phi_{j1}^-(\tilde{n})}{\lambda_j^- \left\|\vec{\Phi}_j^-\right\|^2} - \sum_{j=0}^{\infty} \frac{\Phi_{j2}^+(1)\Phi_{j1}^+(\tilde{n})}{\lambda_j^+ \left\|\vec{\Phi}_j^+\right\|^2} = 1 \tag{5.42}$$

This expression can be derived in the following way. We start by expanding a vector $\vec{f} = (1,0)^T$ according to Eq. (5.29). This results in

$$\begin{pmatrix} 1 \\ 0 \end{pmatrix} = \sum_{j=0}^{\infty} \frac{\left\langle \vec{f}, \vec{\Phi}_j \right\rangle}{\left\|\vec{\Phi}_j\right\|^2} \vec{\Phi}_j(\tilde{n}) \tag{D.17}$$

The expression $\left\langle \vec{f}, \vec{\Phi}_j \right\rangle$ can now be evaluated from Eq. (5.14) which results in

$$\left\langle \vec{f}, \vec{\Phi}_j \right\rangle = \int_0^1 \left[\frac{a_1(\tilde{n})\tilde{r}^F}{\mathrm{Pe}_L^2} \cdot 1 \cdot \Phi_{j1}(\tilde{n}) \right] d\tilde{n} \tag{D.18}$$

From this equation, one obtains for the first vector component of Eq. (D.17)

$$1 = \sum_{j=0}^{\infty} \frac{\Phi_{j1}}{\left\|\vec{\Phi}_j\right\|^2} \int_0^1 \frac{a_1(\tilde{n})\tilde{r}^F \Phi_{j1}}{\mathrm{Pe}_L^2} d\tilde{n} \tag{D.19}$$

Now the integral in Eq. (D.19) can be rewritten. First, one can replace the integrand by using Eq. (5.21). This results in

$$\int_0^1 \frac{a_1(\tilde{n})\tilde{r}^F \Phi_{j1}}{\mathrm{Pe}_L^2} d\tilde{n} = \frac{1}{\lambda_j}\left\{ \int_0^1 \tilde{u}\tilde{r}^F \Phi_{j1} d\tilde{n} - \Phi_{j2}(1) \right\} \tag{D.20}$$

The integral on the right hand side of Eq. (D.20) can be expressed as

$$\frac{1}{\lambda_j} \int_0^1 \tilde{u}\tilde{r}^F \Phi_{j1} d\tilde{n} = \frac{1}{\lambda_j}\left(\Phi_{j1} \int_0^{\tilde{n}} \tilde{u}\tilde{r}^F d\tilde{n} \right)\Bigg|_0^1 - \frac{1}{\lambda_j} \int_0^1 \Phi_{j1}'\left(\int_0^{\tilde{n}} \tilde{u}\tilde{r}^F d\hat{n} \right) d\tilde{n} \tag{D.21}$$

Evaluating the first expression on the right hand side of Eq. (D.21) shows that this term is zero. If one further replaces Φ_{j1}' by Eq. (5.22), one obtains

$$-\frac{1}{\lambda_j} \int_0^1 \Phi_{j1}'\left(\int_0^{\tilde{n}} \tilde{u}\tilde{r}^F d\hat{n} \right) d\tilde{n} = -\int_0^1 \frac{\Phi_{j2}}{\tilde{r}^F a_2}\left(\int_0^{\tilde{n}} \tilde{u}\tilde{r}^F d\hat{n} \right) d\tilde{n} = -\int_0^1 \Phi_{j2}\, \gamma(\tilde{n})\, d\tilde{n} \tag{D.22}$$

If one replaces now the integral in Eq. (D.19) one obtains

$$1 = -\sum_{j=0}^{\infty} \frac{\Phi_{j1}(\tilde{n})}{\left\|\vec{\Phi}_j\right\|^2} \int_0^1 \Phi_{j2} \, \gamma(\tilde{n}) \, d\tilde{n} - \sum_{j=0}^{\infty} \frac{\Phi_{j1}(\tilde{n})\Phi_{j2}(1)}{\left\|\vec{\Phi}_j\right\|^2} \tag{D.23}$$

In order to show that this expression is identical to Eq. (5.42), one has to show that the first sum on the right hand side of this equation is zero. Expanding the vector $\vec{f} = \left(0, \gamma(\tilde{n})\tilde{r}^F a_2\right)^T$ into a series, one obtains for the first vector component

$$0 = \sum_{j=0}^{\infty} \frac{\Phi_{j1}}{\left\|\vec{\Phi}_j\right\|^2} \int_0^1 \gamma(\tilde{n}) \Phi_{j2} \, d\tilde{n} \tag{D.24}$$

Eq. (D.24) shows that the first sum in Eq. (D.23) is zero and one obtains therefore

$$1 = -\sum_{j=0}^{\infty} \frac{\Phi_{j1}(\tilde{n})\Phi_{j2}(1)}{\left\|\vec{\Phi}_j\right\|^2} \tag{D.25}$$

If we distinguish explicitly between positive and negative eigenfunctions in Eq. (D.25), we obtain Eq. (5.42).

D.5 Simplification of the Expression for the Temperature Distribution (for Constant Wall Heat Flux)

For the case of a prescribed wall heat flux, we can simplify Eq. (5.66) by replacing the first two terms in this equation. In the following section, it will be explained how this can be done. The first term in Eq. (5.66) to be replaced is

$$\sum_{j=0}^{\infty} \frac{\Phi_{j1}(1)}{\lambda_j^2 \left\|\vec{\Phi}_j\right\|^2} \Phi_{j1}(\tilde{n}) \tag{D.26}$$

We start our considerations by expanding the vector $\vec{f} = (1,0)^T$ into a series. For the first vector component one obtains

$$1 = \sum_{j=0}^{\infty} \frac{\Phi_{j1}}{\left\|\vec{\Phi}_j\right\|^2} \int_0^1 \frac{a_1(\tilde{n})\tilde{r}^F \Phi_{j1}}{Pe_L^2} \, d\tilde{n} \tag{D.19}$$

The integral in the above equation can be rewritten by using Eq. (5.23). This results in

$$\int_0^1 \frac{a_1(\tilde{n})\tilde{r}^F \Phi_{j1}}{Pe_L^2} d\tilde{n} = -\frac{1}{\lambda_j^2} \int_0^1 \frac{d}{d\tilde{n}}\left[\tilde{r}^F a_2(\tilde{n})\Phi'_{j1}\right] d\tilde{n} + \frac{1}{\lambda_j} \int_0^1 \tilde{u}\tilde{r}^F \Phi_{j1} d\tilde{n} \tag{D.27}$$

Carrying out the integrations in the above equation, it can be seen that the first integral is zero for the case of a prescribed heat flux at the wall and one obtains

$$1 = \sum_{j=0}^{\infty} \frac{\Phi_{j1}(\tilde{n})}{\lambda_j \left\| \vec{\Phi}_i \right\|^2} \int_0^1 \tilde{u} \tilde{r}^F \Phi_{j1}(\tilde{n}) d\tilde{n} \tag{D.28}$$

The integral in Eq. (D.28) can be further rewritten by partial integration. This results in

$$1 = \sum_{j=0}^{\infty} \frac{\Phi_{j1}(\tilde{n})}{\lambda_j \left\| \vec{\Phi}_i \right\|^2} \frac{\Phi_{j1}(1)}{F+1} - \sum_{j=0}^{\infty} \frac{\Phi_{j1}(\tilde{n})}{\lambda_j \left\| \vec{\Phi}_i \right\|^2} \int_0^1 \Phi_{j2}(\tilde{n}) \omega(\tilde{n}) \; d\tilde{n} \tag{D.29}$$

$$\omega(\tilde{n}) = \frac{1}{\tilde{r}^F a_2(\tilde{n})} \int_0^{\tilde{n}} \tilde{u} \tilde{r}^F d\tilde{n}$$

In order to obtain an expression for Eq. (D.26), the second term on the right hand side of Eq. (D.29) needs to be evaluated. Expanding the vector $\vec{f} = \left(0, \omega(\tilde{n}) \tilde{r}^F a_2(\tilde{n})\right)^T$ into a series results in

$$\begin{pmatrix} 0 \\ \omega(\tilde{n}) \tilde{r}^F a_2(\tilde{n}) \end{pmatrix} = \sum_{j=0}^{\infty} \frac{\left\langle \vec{f}, \vec{\Phi}_j \right\rangle}{\left\| \vec{\Phi}_j \right\|^2} \vec{\Phi}_j(\tilde{n}) = \sum_{j=0}^{\infty} \frac{\vec{\Phi}_j(\tilde{n})}{\left\| \vec{\Phi}_j \right\|^2} \int_0^1 \omega(\tilde{n}) \Phi_{j2} d\tilde{n} \tag{D.30}$$

Taking the first vector component of Eq. (D.30) shows that the second term on the right hand side of Eq. (D.29) is zero and we obtain for the expression given by Eq. (D.26)

$$\sum_{j=0}^{\infty} \frac{\Phi_{j1}(1)}{\lambda_j \left\| \vec{\Phi}_j \right\|^2} \Phi_{j1}(\tilde{n}) = F+1 \tag{D.31}$$

In addition, we want to derive an analytical expression for the first sum on the right hand side of Eq. (5.66):

$$\sum_{j=0}^{\infty} \frac{\Phi_{j1}(1)}{\lambda_j^2 \left\| \vec{\Phi}_j \right\|^2} \Phi_{j1}(\tilde{n}) \tag{D.32}$$

If we put $\tilde{x} \to \infty$ in Eq. (5.66), we obtain the fully developed temperature distribution in the duct with a prescribed heat flux at the wall. Using Eq. (D.31) this distribution can be written as

$$\tilde{x} \to \infty : \Theta_\infty(\tilde{x}, \tilde{n}) = \sum_{j=0}^{\infty} \frac{\Phi_{j1}(1)}{\lambda_j^2 \left\| \vec{\Phi}_j \right\|^2} \Phi_{j1}(\tilde{n}) + \tilde{x}(F+1) \tag{D.33}$$

As it can be seen from Eq. (D.33) this temperature distribution contains a term, which is linearly increasing in $\tilde{x}$ and a term, which is a function of $\tilde{n}$. Therefore, Eq. (D.33) might be written as

$$\tilde{x} \to \infty : \quad \Theta_\infty(\tilde{x}, \tilde{n}) = \Psi(\tilde{n}) + \tilde{x}(F+1) \tag{D.34}$$

The function $\Psi(\tilde{n})$, appearing in Eq. (D.34), is not known. However, this function can easily been obtained by inserting Eq. (D.34) into the energy equation (5.6) and solving the resulting ordinary differential equation for $\Psi(\tilde{n})$. One finally gets

$$\Psi(\tilde{n}) = \int_0^{\tilde{n}} \frac{F+1}{\tilde{r}^F a_2(\bar{n})} \int_0^{\bar{n}} \tilde{u}(s) \tilde{r}^F \, ds \, d\bar{n} + C_2 = \overline{\Psi}(\tilde{n}) + C_2 \tag{D.35}$$

The constant C_2, which appears in the above equation, can be derived from a global energy balance. This might be done by evaluating Eq. (5.9) for the case of a thermally fully developed flow $(\tilde{x} \to \infty)$ for the whole flow domain $E(\tilde{n} = 1, \tilde{x} \to \infty)$. By considering the boundary conditions for the function E, given by Eq. (5.58), one obtains

$$\tilde{x} \to \infty : E(1, \tilde{x}) = \tilde{x} = \int_0^1 \left[\tilde{u}\Theta_\infty - \frac{1}{\mathrm{Pe}_L^2} a_1(\bar{n}) \frac{\partial \Theta_\infty}{\partial \tilde{x}} \right] \tilde{r}^F \, d\tilde{n} \tag{D.36}$$

Inserting the temperature distribution given by Eq. (D.34) into Eq. (D.36) results in

$$\int_0^1 \tilde{u} \, \tilde{n}^F \Psi(\tilde{n}) \, d\tilde{n} = \frac{(F+1)}{\mathrm{Pe}_L^2} \int_0^1 a_1(\tilde{n}) \, \tilde{r}^F \, d\tilde{n} \tag{D.37}$$

Comparing Eq. (D.37) with Eq. (D.35) results in the following expression for the constant C_2

$$C_2 = \frac{(F+1)^2}{\mathrm{Pe}_L^2} \int_0^1 a_1(\tilde{n}) \, \tilde{r}^F \, d\tilde{n} - (F+1) \int_0^1 \tilde{u} \, \tilde{r}^F \overline{\Psi}(\tilde{n}) \, d\tilde{n} \tag{D.38}$$

$$\overline{\Psi}(\tilde{n}) = \int_0^{\tilde{n}} \frac{F+1}{\tilde{r}^F a_2(\tilde{n})} \int_0^{\bar{n}} \tilde{u} \, \tilde{r}^F \, ds \, d\bar{n} \tag{D.39}$$

From this equation one finally obtains for the sum, Eq. (D.32),

$$\sum_{j=0}^{\infty} \frac{\Phi_{j1}(1)}{\lambda_j^2 \|\vec{\Phi}_j\|^2} \Phi_{j1}(\tilde{n}) = \Psi(\tilde{n}) = \overline{\Psi}(\tilde{n}) + C_2 \tag{D.40}$$

It should be noted here that the above analytical expressions, Eqs. (D.26) and (D.32), transform into the expressions developed by Papoutsakis et al (1980), if we consider the simplified case of a laminar pipe flow ($F = 1$, $a_1 = a_2 = 1$, $\tilde{u} = 2(1-\tilde{n}^2)$). For this case one obtains

$$\sum_{j=0}^{\infty} \frac{\Phi_{j1}(\tilde{n})\Phi_{j1}(1)}{\lambda_j \left\| \vec{\Phi}_j \right\|^2} = 2 \tag{D.41}$$

and

$$\sum_{j=0}^{\infty} \frac{\Phi_{j1}(1)\Phi_{j1}(\tilde{n})}{\lambda_j^2 \left\| \vec{\Phi}_j \right\|^2} = \tilde{n}^2 - \frac{\tilde{n}^4}{4} + \frac{8}{Pe_D^2} - \frac{7}{24} \tag{D.42}$$

These expressions have also been given in Eq. (5.75). For the case of a planar channel, Eq. (5.76) is obtained.

D.6 The Vector Norm $\left\| \vec{\Phi}_j \right\|^2$

In this section, some interesting results about the vector norm $\left\| \vec{\Phi}_j \right\|^2$ are presented. The vector norm is defined by inserting the eigenvector into Eq. (5.14). This results in

$$\left\| \vec{\Phi}_j \right\|^2 = \left\langle \vec{\Phi}_j, \vec{\Phi}_j \right\rangle = \int_0^1 \left[\frac{a_1(\tilde{n})\tilde{r}^F}{Pe_L^2} \Phi_{j1}^2(\tilde{n}) + \frac{\Phi_{j2}^2(\tilde{n})}{a_2(\tilde{n})\tilde{r}^F} \right] d\tilde{n} \tag{D.43}$$

First, we establish a connection between the vector norm and the derivative of the eigenfunction with respect to the eigenvalue (see Eq. (5.46) and Eq. (5.71)). This connection is important because it allows us to evaluate the constants in the temperature distribution very easily. Therefore, we consider two vectors $\vec{\Phi}_j$ and $\vec{\Phi}$ which both satisfy Eq. (5.20). This results in

$$\underset{\sim}{L}\vec{\Phi}_j = \lambda_j \vec{\Phi}_j \tag{D.44}$$
$$\underset{\sim}{L}\vec{\Phi} = \lambda \vec{\Phi}$$

The eigenvector $\vec{\Phi}_j$ satisfies both boundary conditions at $\tilde{n}=0$ and $\tilde{n}=1$, whereas the vector $\vec{\Phi}$ satisfies the boundary condition at $\tilde{n}=0$ and only in the limit $\lambda \to \lambda_j$ the boundary condition at $\tilde{n}=1$.

If we now apply the inner product, given by Eq. (5.14), to Eq. (D.44), we obtain

$$\left\langle \underline{L}\vec{\Phi}_j, \vec{\Phi} \right\rangle - \left\langle \vec{\Phi}_j, \underline{L}\vec{\Phi} \right\rangle = \left(\lambda_j - \lambda \right) \left\langle \vec{\Phi}_j, \vec{\Phi} \right\rangle \qquad (D.45)$$

The expression on the left hand side of Eq. (D.45) can be evaluated to be

$$\left\langle \underline{L}\vec{\Phi}_j, \vec{\Phi} \right\rangle - \left\langle \vec{\Phi}_j, \underline{L}\vec{\Phi} \right\rangle = \left[\Phi_{j1}\Phi_2 - \Phi_{j2}\Phi_1 \right]_0^1 \qquad (D.46)$$

Because $\Phi_{j2}(0) = \Phi_2(0) = 0$ one obtains from Eq. (D.45)

$$\left\langle \vec{\Phi}_j, \vec{\Phi} \right\rangle = \frac{\Phi_{j1}(1)\Phi_2(1,\lambda) - \Phi_{j2}(1)\Phi_1(1,\lambda)}{\lambda_j - \lambda} \qquad (D.47)$$

If we evaluate Eq. (D.47) for the limiting case when $\lambda \to \lambda_j$, $\vec{\Phi} \to \vec{\Phi}_j$ we obtain

$$\lim_{\lambda \to \lambda_j} \left\langle \vec{\Phi}_j, \vec{\Phi} \right\rangle = \left\| \vec{\Phi}_j \right\|^2 = \lim_{\lambda \to \lambda_j} \frac{\Phi_{j1}(1)\Phi_2(1,\lambda) - \Phi_{j2}(1)\Phi_1(1,\lambda)}{\lambda_j - \lambda} \qquad (D.48)$$

Before evaluating the expression on the right hand side of Eq. (D.48), the wall boundary conditions for $\tilde{n} = 1$ need to be specified.

Constant Wall Temperature: $\Phi_{j1}(1) = 0$

Inserting this boundary condition into Eq. (D.48) one obtains

$$\lim_{\lambda \to \lambda_j} \left\langle \vec{\Phi}_j, \vec{\Phi} \right\rangle = \left\| \vec{\Phi}_j \right\|^2 = \lim_{\lambda \to \lambda_j} \left\{ -\Phi_{j2}(1) \frac{\Phi_1(1,\lambda)}{\lambda_j - \lambda} \right\} \qquad (D.49)$$

If $\lambda \to \lambda_j$ is introduced into the Eq. (D.49), the expression on the right hand side leads to 0/0 and can be evaluated by the rule of L'Hospital. One obtains Eq. (5.46)

$$\left\| \vec{\Phi}_j \right\|^2 = -\Phi_{j2}(1) \frac{d\Phi_{j1}(1,\lambda)}{d\lambda} \bigg|_{\lambda = \lambda_j} \qquad (5.46)$$

Constant Wall Heat Flux: $\Phi_{j2}(1) = 0$

For this case one obtains from Eq. (D.48)

$$\lim_{\lambda \to \lambda_j} \left\langle \vec{\Phi}_j, \vec{\Phi} \right\rangle = \left\| \vec{\Phi}_j \right\|^2 = \lim_{\lambda \to \lambda_j} \frac{\Phi_{j1}(1)\Phi_2(1,\lambda)}{\lambda_j - \lambda} \qquad (D.50)$$

Using the rule of L'Hospital to evaluate the term on the right hand side results in

$$\left\| \vec{\Phi}_j \right\|^2 = \Phi_{j1}(1) \frac{d\Phi_{j2}(1,\lambda)}{d\lambda} \bigg|_{\lambda = \lambda_j} \qquad (D.51)$$

In Eq. (D.51) the vector component Φ_2 can be replaced by using Eq. (5.22) at the wall. This leads finally to

$$\left\|\vec{\Phi}_j\right\|^2 = \Phi_{j1}(1)\frac{d}{d\lambda}\left(\frac{\Phi'_{j1}(1,\lambda)}{\lambda}\right)\Bigg|_{\lambda=\lambda_j} \tag{D.52}$$

Finally, for the case of a constant wall temperature, Eq. (5.45) is derived

$$\left\|\vec{\Phi}_j\right\|^2 = -\frac{1}{\lambda_j}\int_0^1 \tilde{r}^F\tilde{u}\,\Phi^2_{j1}d\tilde{n} + \frac{2}{Pe_L^2}\int_0^1 \tilde{r}^F a_1(\tilde{n})\Phi^2_{j1}d\tilde{n} \tag{5.45}$$

In order to derive this equation, we start from Eq. (D.43)

$$\left\|\vec{\Phi}_j\right\|^2 = \left\langle\vec{\Phi}_j,\vec{\Phi}_j\right\rangle = \int_0^1\left[\frac{a_1(\tilde{n})\tilde{r}^F}{Pe_L^2}\Phi^2_{j1}(\tilde{n}) + \frac{\Phi^2_{j2}(\tilde{n})}{a_2(\tilde{n})\tilde{r}^F}\right]d\tilde{n} \tag{D.43}$$

For Eq. (D.43), one can rewrite the second term in the integral by using Eq. (5.22). This results in

$$\int_0^1 \frac{\Phi^2_{j2}(\tilde{n})}{a_2(\tilde{n})\tilde{r}^F}\,d\tilde{n} = \frac{1}{\lambda_j}\int_0^1 \Phi_{j2}\Phi'_{j1}d\tilde{n} \tag{D.53}$$

The first term in the integral of Eq. (D.43) can be rewritten by using Eq. (5.21)

$$\int_0^1 \frac{a_1(\tilde{n})\tilde{r}^F}{Pe_L^2}\Phi^2_{j1}(\tilde{n})\,d\tilde{n} = \frac{1}{\lambda_j}\int_0^1\left(\tilde{u}\tilde{r}^F\Phi^2_{j1} - \Phi_{j1}\Phi'_{j2}\right)d\tilde{n} \tag{D.54}$$

Introducing Eqs. (D.53-D.54) into Eq. (D.43) gives

$$\left\|\vec{\Phi}_j\right\|^2 = \frac{1}{\lambda_j}\int_0^1\left[\tilde{u}\tilde{r}^F\,\Phi^2_{j1}(\tilde{n}) + \Phi'_{j1}\Phi_{j2} - \Phi_{j1}\Phi'_{j2}\right]d\tilde{n} \tag{D.55}$$

and after partial integration (with the boundary conditions $\Phi_{j2}(0) = \Phi_{j1}(1) = 0$) one obtains from Eq. (D.55)

$$\left\|\vec{\Phi}_j\right\|^2 = \frac{1}{\lambda_j}\int_0^1\tilde{u}\tilde{r}^F\,\Phi^2_{j1}(\tilde{n})d\tilde{n} - \frac{2}{\lambda_j}\int_0^1\Phi_{j1}\Phi'_{j2}\,d\tilde{n} \tag{D.56}$$

The second integral in Eq. (D.56) can now be modified by using Eq. (5.22)

$$-\frac{2}{\lambda_j}\int_0^1\Phi_{j1}\Phi'_{j2}\,d\tilde{n} = -\frac{2}{\lambda_j}\int_0^1\tilde{u}\tilde{r}^F\Phi^2_{j1}d\tilde{n} + \frac{2}{\lambda_j}\int_0^1\frac{a_1\tilde{r}^F}{Pe_L^2}\Phi^2_{j1}d\tilde{n} \tag{D.57}$$

Inserting Eq. (D.57) into Eq. (D.56) results in Eq. (5.45).

References

Abbrecht PH, Churchill SW (1960) The thermal entrance region in fully developed turbulent flow. AICHE Journal 6:268 - 273.

Acrivos A (1980) The extended Graetz problem at low Peclet numbers. Appl. Sci. Res. 36:35 - 40.

Adams WH Mc (1954) Heat Transmission. Mc Graw Hill, New York.

Agrawal HC (1960) Heat Transfer in laminar flow between parallel plates at small Peclet numbers. Appl. Sci. Res. 9:177 - 189.

Akyildiz FT, Bellout H (2004) The extended Graetz problem for dipolar fluids. Int. J. Heat Mass Transfer 47: 2747-2753.

Ames WF (1965a) Nonlinear partial differential equations in Engineering. Academic Press, New York.

Ames WF (1965b) Similarity for the nonlinear diffusion equation. I&EC Fundamentals 4: 72 – 76.

Andrews LC, Shivamoggi BK (1988) Integral Transforms for Engineers and Applied Mathematicians, Macmillian Publ. Comp., New York.

Antonia RA, Kim J (1991) Reynolds shear stress and heat flux calculations in a fully developed turbulent duct flow. Int. J. Heat Mass Transfer 34:2013- 2018.

Awad AS (1965) Heat transfer and eddy diffusivity in NaK in a pipe at uniform wall temperature. Ph.D. Thesis, University of Washington, Seattle.

Azer NZ, Chao BT (1960) A mechanism of turbulent heat transfer in liquid metals. Int. J. Heat Mass Transfer 1:121 - 138.

Aziz A, Na TY (1984) Perturbation Methods in Heat Transfer. Springer, Berlin, Heidelberg New York.

Batchelor GK (1949) Diffusion in a field of homogeneous turbulence. Austral. J. Sci. Res. A2: 437 – 450.

Bayazitoglu Y, Özisik MN (1980) On the solution of Graetz type problems with axial conduction. Int. J. Heat Mass Transfer 23:1399 - 1402.

Bhatti MS, Shah RK (1987) Turbulent and transition flow convective heat transfer in ducts. Kakaç S; Shah RK, Aung W (eds): Handbook of single-phase convective heat transfer, John Wiley and sons, New York.

Bilir S (1992) Numerical solution of Graetz problem with axial conduction. Numerical Heat Transfer, Part A 21: 493 – 500.

Birkhoff G (1948) Dimensional analysis of partial differential equations. Electrical Eng. 67: 1185 – 1188.

Borisenko AI, Kostinov ON, Chumachenko VI (1973) Experimental study of turbulent flow in a rotating channel. J. of Eng. Physics 24: 770 – 773.

Bowman F (1958) Introduction to Bessel functions. Dover Publ., New York.

Bleustein JL, Green AE (1967) Dipolar fluids. Int. J. Eng. Sci. 5: 323 - 340

Bradshaw P (1969) The analogy between streamline curvature and buoyancy in turbulent shear flow. J. Fluid Mech. 36:177 - 191.

Broman A (1970) Introduction to partial differential equations. Dover Publications, INC., New York.

Brown GM (1960) Heat or mass transfer in a fluid in laminar flow in a circular or flat conduit. AICHE Journal 6:179 - 183.

Bronstein IN, Semendjajew KA (1981) Taschenbuch der Mathematik. Verlag Harri Deutsch, Frankfurt.

Burow P, Weigand B (1990) One dimensional heat conduction is a semi-infinite solid with the surface temperature a harmonic function of time: A simple approximate solution for the transient behavior. ASME Journal of Heat Transfer 112: 1076-1079.

Carslaw HS, Jaeger, J.C. (1992) Conduction of Heat in Solids, Oxford University Press.

Cebeci T, Bradshaw P (1984) Physical and Computational Aspects of Convective Heat Transfer. Springer, New York.

Cebeci T, Chang KC (1978) A general method for calculating momentum and heat transfer in laminar and turbulent duct flow. Num. Heat Transfer 1:39 - 68.

Cochran WG (1934) The flow due to a rotating disk. Proc. Cambr. Phil. Soc. 30: 365 – 375.

Coddington EA, Levinson N (1955) Theory of ordinary differential equations. Mc Graw Hill, New York.

Cohen CB, Reshotko E (1956) The compressible laminar boundary layer with heat transfer and arbitrary pressure gradient. NACA Rep. Nr. 1294.

Collatz L (1981) Differentialgleichungen. B.G. Teubner, Stuttgart.

Courant R, Hilbert D (1991) Methoden der mathematischen Physik. Springer, Berlin Heidelberg New York.

Chieng CC, Launder BE (1980) On the calculation of turbulent heat transfer downstream from an abrupt pipe expansion, Num. Heat Transfer 3:189-207.

Darcy H (1858) Récherches expérimentales relatives aux mouvements de l'eau dans tuyaux. Mem. prés à l'Acedémie des Sciences de l'Institute de France 15: 141.

Deavours CA (1974) An exact solution for the temperature distribution in parallel plate Poiseuille flow. Journal of Heat Transfer 96:489 - 495.

Dewey CF, Gross JF (1967) Exact similar solutions of the laminar boundary – Layer equations. Adv. Heat Transfer 4: 317 – 446.

Dhawan S (1953) Direct measurements of skin friction. NACA Tech. Rep. 1121.

Deissler RG (1955) Analysis of turbulent heat transfer, mass transfer and friction in smooth tubes at high Prandtl and Schmidt numbers. NACA Tech. Rep. 1210.

Driest ER van (1951) Turbulent boundary layers in compressible fluids. J. of the Aeron. Sci. 18:145 - 160.

Driest ER van (1956) On turbulent flow near a wall. J. of Aeron. Sci. 23: 1007 – 1011.

Dwyer OE (1965) Heat transfer to liquid metals flowing turbulently between parallel plates. Nucl. Sci. Eng. 21:79 - 89.

Dyke MD van (1964) Perturbation methods in fluid mechanics, Academic Press, New York.

Ebadian MA, Zhang HY (1989) An exact solution of extended Graetz problem with axial heat conduction. Int. J. Heat Mass Transfer 32:1709 - 1717.

Eckert ERG, Drake M (1987) Analysis of heat and mass transfer. Hemisphere Publ. Corp., New York.

Ede AJ (1967) Advances in free convection. Adv. Heat Transfer 4:1 – 64.

Eggels JGM, Nieuwstadt FTM (1993) Large eddy simulation of turbulent flow in an axially rotating pipe. Proc. of the 9th Symp. on Turbulent Shear Flows, August 16 - 18, Kyoto, Japan, pp 1 - 4.

Erdelyi A (1956) Asymptotic Expansions. Dover Publ., New York.

Faggiani S, Gori F (1980) Influence of streamwise molecular heat conduction on the heat transfer for liquid metals in turbulent flow between parallel plates. Journal of Heat Transfer 102:292 - 296.

Froman N, Froman PO (1965) JWKB Approximation, Contributions to the Theory. North-Holland, Amsterdam.

Fuchs H (1973) Wärmeübergang an strömendes Natrium. Thesis, Eidg. Institut für Reaktorforschung, Würenlingen, Switzerland, EIR-Bericht 241.

Fuller RE, Samuels MR (1971) Simultaneous development of the velocity and temperature fields in the entry region of an annulus, Chem. Eng. Prog. Symp. Ser. 113:71 - 77.

Gibson MM, Younis BA (1986) Caculation of swirling jets with a Reynolds-stress closure, Physics of Fluids 29:38.

Gilliland ER; Musser RJ, Page WR (1951) Heat transfer to Mercury. Gen. Disc. on Heat Transfer, Inst. Mech. Eng. and ASME, London, pp. 402 – 404.

Görtler (1975) Dimensionsanalyse: Theorie der physikalischen Dimensionen mit Anwendungen. Spinger Verlag, Berlin.

Graetz L (1883) Über die Wärmeleitfähigkeit von Flüssigkeiten, Teil 1. Ann. Phys. Chem. 18:79 - 84.

Graetz L (1885) Über die Wärmeleitfähigkeit von Flüssigkeiten, Teil 2. Ann. Phys. Chem. 25:337 -357.

Gradshteyn IS, Ryzhik IM (1994) Table of Integrals, Series and Products. Academic Press, New York.

Hansen AG (1964) Similarity analyses of boundary value problems in Engineering. Prentice-Hall.

Hennecke D (1968) Heat Transfer by Hagen-Poiseuille flow in the thermal development region with axial conduction. Wärme- und Stoffübertragung 1:177 - 184.

Hirai S, Takagi T (1987) Prediction of heat transfer deterioration in turbulent swirling pipe flow. Proc. of the 2nd ASME/JSME Thermal Eng. Joint Conf., Vol. 5, pp.181 - 187.

Hirai S, Takagi T, Matsumoto M (1988) Predictions of the laminarization phenomena in an axially rotating pipe flow. J. of Fluids, Eng 110:424 –430.

Hsu CJ (1965) Heat transfer in a round tube with sinusoidal wall flux distribution. AICHE Journal 11:690.

Hsu CJ (1970) Theoretical solutions for low Peclet number thermal-entry-region heat transfer in laminar flow through concentric annuli. Int. J. Heat Mass Transfer 13:1907 - 1924.

Hsu CJ (1971a) Laminar flow heat transfer in circular or parallel-plate channels with internal heat generation and the boundary conditions of the third kind. J. of the Chin. Inst. of Chem. Eng., pp. 85 - 100.

Hsu CJ (1971b) An exact analysis of low Peclet number thermal entry region heat transfer in transversely nonuniform velocity fields, AICHE Journal 17:732 - 740.

Jaluria Y (1980) Natural convection heat and mass transfer. Pergamon Press, Oxford.

Jischa M, Rieke HB (1979) About the prediction of turbulent Prandtl and Schmidt numbers. Int. J. Heat Mass Transfer 24: 1179 – 1189.

Jischa M (1982) Konvektiver Impuls-, Wärme- und Stoffaustausch., Vieweg & Sohn, Braunschweig.

Jones AS (1971) Extensions to the solution of the Graetz problem. Int. J. Heat Mass Transfer 14:619 - 623.

Kader BA (1971) Heat and mass transfer in laminar flow in the entrance section of a circular tube. High Temp. (USSR) 9:1115 - 1120.

Kamke E (1983) Differentialgleichungen: Lösungsmethoden und Lösungen, Band 1. B.G Teubner, Stuttgart.

Karman T von (1921) Über laminare und turbulente Reibung. ZAMM 1: 233 – 252.

Karman T von (1939) The analogy between fluid friction and heat transfer, Trans. ASME 61:705 - 710.

Kays WM, Crawford ME (1993) Convective Heat and Mass Transfer, Mc. Graw-Hill Inc., New York.

Kays WM (1994) Turbulent Prandtl number - Where are we? Journal of Heat Transfer 116:285-295.

Kevorkian J, Cole JD (1981) Perturbation Methods in Applied Mathematics. Appl. Math. Sci., Vol. 34, Springer, Berlin.

Kikuyama K, Murakami M, Nishibori K, Maeda K (1983) Flow in an axially Rotating pipe. Bull. JSME 26:506- 513.

Koosinlin ML, Launder BE, Sharma BI (1975) Prediction of momentum, heat and mass transfer in swirling, turbulent boundary layers. Journal of Heat Transfer 96:204 - 209.

Kumar IJ (1972) Recent mathematical methods in heat transfer. Adv. Heat Transfer, Vol. 8, pp. 1 - 91.

Latzko H (1921) Der Wärmeübergang an einem turbulenten Flüssigkeits- oder Gasstrom. ZAMM 1:268 - 290.

Laufer J (1950) Some recent measurements in a two-dimensional turbulent channel. Journal Aerosp. Sci. 17:277 - 280.

Lauffer D (2003) Wärmeübertragung im laminar oder turbulent durchströmten Rohr und ebenen Kanal bei kleinen Prandtlzahlen. Diploma-Thesis, Stuttgart University, Germany.

Launder BE (1988) On the computation of convective heat transfer in complex turbulent flow. Journal of Heat Transfer 110:1112-1128.

Lauwerier HA (1950) The use of confluent hypergeometric functions in mathematical Physics and the solution of an eigenvalue problem. Appl. Sci. Res., Sect. A2:184 - 204.

Lawn CJ (1969) Turbulent heat transfer at low Reynolds number. Journal of Heat Transfer 91:532.

Lee SL (1982) Forced convection heat transfer in low Prandtl number turbulent flows: influence of axial conduction. Can. J. Chem. Eng. 60:482-486.

Leveque MA (1928) Les lois de la transmission de chaleur par convection. Ann. Mines. Mem. 13:201 - 415.

Levy F (1929) Strömungserscheinungen in rotierenden Rohren. VDI Forschungsarbeiten auf dem Gebiet des Ingenieurwesens 322:18 - 45.

Levy S (1952) Heat transfer to constant-property laminar boundary-layer flows with power-function free-stream velocity and wall temperature variation. J. Aeron. Sci. 19:341 - 348.

Li TY, Nagamatsu HT(1953) Similar solutions of compressible boundary layer equations. J. of Aeron. Sci. 20:653 - 656.

Li TY, Nagamatsu HT (1955) Similar solutions of compressible boundary layer equations. J. of Aeron. Sci. 22:607 - 616.

Liouville J (1837) Journal de Math. 2:24.

Loitsianski LG (1967) Laminare Grenzschichten. Akademie-Verlag, Berlin.

Ludwieg H (1956) Bestimmung des Verhältnisses der Austauschkoeffizienten für Wärme- und Impuls bei turbulenten Grenzschichten. Zeitschrift für Flugwissenschaften 4:73 - 81.

Lundberg RE; Mc Cuen PA, Reynolds WC(1963): Heat transfer in annular passages. Hydrodynamically developed laminar flow with arbitrarily prescribed wall temperatures or heat fluxes. Int. J. Heat Mass Transfer 6:495 – 529.

Mackrodt PA (1971) Stabilität von Hagen-Poiseuille-Strömungen mit überlagerter starrer Rotation. Mitteilungen MPI Strömungsforschung und Aerodynamische Versuchsanstalt Göttingen, Vol. 55.

Malin MR, Younis BA (1997) The prediction of turbulent transport in an axially rotating pipe. Int. Com. Heat Mass Transfer 24:89-98.

Mikhailov MD, Özisik MN (1984) Unified Analysis and Solutions of Heat and Mass Diffusion. John Wiley, New York.

Millsaps K, Pohlhausen K (1952) Heat transfer by laminar flow from a rotating plate. Journal Aeron. Sci. 19:120-126.

Mizushina T, Ogino F (1970) Eddy viscosity and universal velocity profile in turbulent flow in a straight pipe. J. of Chem. Eng. of Japan 3:166 - 170.

Morgan AJA (1952) The reduction by one of the number of independent variables in some systems of partial differential equations. Quart. J. Math. Oxford 3: 250 – 259

Müller EA, Matschat K (1962) Über das Auffinden von Ähnlichkeitslösungen partieller Differentialgleichungssysteme unter Benutzung von Transformationsgruppen, mit Anwendungen auf Probleme der Strömungsphysik. Misz. der angew. Mechanik, Akademie Verlag, Berlin, pp. 190 - 222.

Murakami M, Kikuyama K (1980) Turbulent flow in axially rotating pipes. J. Fluids Eng. 102: 97-103.

Myers GE (1987) Analytical methods in conduction heat transfer. Genium Pulishing Corporation, Schenectady.

Myint- UT, Debnath L (1978) Ordinary differential equations. Elsevier North Holland, Inc., New York.

Newman J (1969) The Graetz problem. UCRL 18646, Dep. Chem. Eng., University of California at Berkely.

Nguyen TV (1992) Laminar heat transfer for thermally developing flow in ducts. Int. J. Heat Mass Transfer 35:1733-1741.

Nikuradse J (1932) Gesetzmäßigkeit der turbulenten Strömung in glatten Rohren. Forsch. Arb. Ing. Wesens, Vol. 356.

Nishibori K, Kikuyama K, Murakami M (1987) Laminarization of turbulent flow in the inlet region of an axially rotating pipe. JSME Int. J. 30:255 - 262.

Notter RH, Sleicher CA (1971a) The eddy diffusivity in turbulent boundary layer near a wall. Chem. Eng. Sci. 26:161 - 171.

Notter RH, Sleicher CA (1971b) A solution to the turbulent Graetz problem by matched asymptotic expansions - II The case of uniform wall heat flux. Chem. Eng. Sci. 26:559-565.

Notter RH, Sleicher CA (1972) A solution to the turbulent Graetz problem - III Fully developed and entry region heat transfer rates. Chem. Eng. Sci. 27:2073 - 2093.

Nusselt W (1910) Die Abhängigkeit der Wärmeübergangszahl von der Rohrlänge. VDI Zeitschrift 54:1154 - 1158.

Ostrach S (1953) An analysis of laminar free-convection flow and heat transfer about a flat plate parallel to the direction of the generating body force. NACA Report 1111.

Özisik MN (1968) Boundary value problems of heat conduction. Dover Publications, New York.

Özisik MN; Cotta RM, Kim WS (1989) Heat transfer in turbulent forced convection between parallel-plates. The Can. J. of Chem. Eng. 67:771 - 776.

Orlandi P, Fatica M (1997): Direct simulation of a turbulent pipe rotating along its axis. J. of Fluid Mech. 343: 43-73.

Papoutsakis E, Ramkrishna D, Lim HC (1980a) The extended Graetz problem with Dirichlet wall boundary conditions. Appl. Sci. Res. 36:13 - 34.

Papoutsakis E, Ramkrishna D, Lim HC (1980b) The entended Graetz problem with prescribed wall flux. AICHE Journal 26:779 - 787.

Papoutsakis E, Ramkrishna D (1981) Heat Transfer in a capillary flow emerging from a reservoir. Journal of Heat Transfer 103:429 - 435.

Plaschko P, Brod K (1989) Höhere mathematische Methoden für Ingenieure und Physiker, Springer, Berlin Heidelberg New York.

Prandtl L (1910) Eine Beziehung zwischen Wärmeaustausch und Strömungswiderstand der Flüssigkeit. Z. Physik 11:1072 - 1078.

Prandtl L (1925) Über die ausgebildete Turbulenz, ZAMM 5:136-139.

Prandtl (1935) The mechanics of viscous fluids. Durand WF (ed), Aerodynamic Theory 3:34-208.

Rayleigh Lord (1917) On the dynamics of revolving fluids. Proc. Royal Society A 93:148 – 154.

Ramkrishna D, Amundson NR (1979a) Boundary value problems in transport with mixed and oblique derivative boundary conditions - I. Formulation of equivalent integral equations. Chem. Eng. Sci. 34:301 - 308.

Ramkrishna D, Amundson NR (1979b) Boundary value problems in transport with mixed and oblique derivative boundary conditions - II. Reduction to first order systems. Chem. Eng. Sci. 34:309 - 318.

Rannie WD (1956) Heat transfer in turbulent shear flow. J. Aeron. Sci. 23:485.

Reed CB (1987) Convective heat transfer in liquid metals. In Handbook of Single Phase Convective Heat Transfer eds. Kakac S, Shah RK, Aung W, Chap. 8, Wiley, New York.

Reich G (1988) Strömung und Wärmeübertragung in einem axial rotierenden Rohr. Dissertation, TH Darmstadt, Germany.

Reich G, Beer H (1989) Fluid flow and heat transfer in an axially rotating pipe - I. Effect of rotation on turbulent pipe flow. Int. J. Heat Mass Transfer 32:551 - 562.

Reich G, Weigand B, Beer H (1989) Fluid flow and heat transfer in an axially rotating pipe - II. Effect of rotation on laminar pipe flow. Int. J. Heat Mass Transfer 32:563 - 574.

Reichhardt H (1940) Die Wärmeübertragung in turbulenten Reibungsschichten. ZAMM 20:297.

Reichhardt H (1951) Vollständige Darstellung der turbulenten Geschwindigkeitsverteilung in glatten Leitungen. ZAMM 31:208 - 219.

Reid WT (1980) Sturmian theory for ordinary differential equations. Appl. Math. Sci., Vol. 31, Springer, Berlin Heidelberg New York.

Reynolds WC (1974) Recent advances in the computation of turbulent flows. Adv. in Chem. Eng. 9:193 -246

Reynolds AJ (1975) The prediction of turbulent Prandtl and Schmidt numbers. Int. J. Heat Mass Transfer 18:1055-1069.

Rinck KJ, Beer H (1999) Numerical calculation of the fully developed turbulent flow in an axially rotating pipe with a second-moment closure. J. Fluids Engng 120:274-279.

Rothe T (1994) Der Einfluß der Rotation auf die Strömung und den Wärmetransport im turbulent durchströmten Ringspalt zwischen zwei rotierenden Wellen. Dissertation, TH Darmstadt, Germany.

Rothfus RR, Walker JE, Whan GA (1958) Correlation of local velocities in tubes, annuli and parallel plates. AICHE Journal 4:240 - 245.

Sadeghipour MS, Özisik MN, Mulligan JC (1984) The effect of external convection on freezing in steady turbulent flow in a tube. Int. J. Eng. Sci. 22:135 - 148.

Sagan (1989) Boundary and eigenvalue problems in Mathematical Physics, Dover Publications, New York.

Sakakibara M, Endoh K (1977) Analysis of heat transfer in the entrance region with fully developed turbulent flow between parallel plates. Heat Transfer Japanese Research 6:54 - 61.

Sauer R, Szabó I (1969) Mathematische Hilfsmittel des Ingenieurs, Teil 1 - 4. Springer, Berlin.

Schlichting H (1982) Grenzschicht-Theorie, G. Braun, Karlsruhe.

Schmidt E, Bechmann W (1930) Das Temperatur- und Geschwindigkeitsfeld von einer Wärme abgebenden senkrechten Platte bei natürlicher Konvektion. Forsch. Ing.-Wes. 1: 391.

Schneider W (1978) Mathematische Methoden der Strömungsmechanik. Vieweg, Braunschweig.

Sellars JR, Tribus M, Klein JS (1956) Heat transfer to laminar flow in a round tube or flat conduit - The Graetz problem extended. Trans. ASME 78 :441 - 448.

Shah RK (1975) Thermal entry length solutions for the circular tube and parallel plates. Proc. Nat. Heat Mass Transfer Conf., 3rd., Indian Inst. Techn., Bombay, Vol. 1, HMT-11-75.

Shah RK, London AL (1978) Laminar flow forced convection in ducts. Adv. in Heat Transfer, Academic Press, New York.

Shchukin VK (1967) Hydraulic resistance of rotating tubes. J. of Eng. Physics 12: 418 –422.

Shibani AA, Özisik MN (1977a) Freezing of liquids in turbulent flow inside tubes. The Can. J. of Chem. Eng. 55:672 - 677.

Shibani AA, Özisik MN (1977b) A solution to heat transfer in turbulent flow between parallel plates. Int. J. Heat Mass Transfer 20:565 - 573.

Simon V (2004) Dimensionsanalyse. Lecture Notes, Institute of Aerospace Thermodynamics, University of Stuttgart, Germany

Simmonds JG, Mann JE (1986) A first look at perturbation theory. R.E. Krieger Publ. Comp., Florida.

Singh SN (1958) Heat Transfer by laminar flow in a cylindrical tube. Appl. Sci. Res., Ser. A 7:325 - 340.

Sleicher CA, Tribus M (1957) Heat Transfer in a pipe with turbulent flow and arbitrary wall-temperature distribution. Trans ASME 79:789 - 797.

Sleicher CA, Notter RH, Crippen MD (1970) A solution to the turbulent Graetz problem by matched asymptotic expansions - I. The case of uniform wall temperature. Chem. Eng. Sci. 25:845 - 857.

Sleicher CA, Awad AS, Notter RH (1973) Temperature and eddy diffusivity profiles in NaK. Int. J. Heat Mass Transfer 16: 1565 – 1575.

Sleicher CA, Rouse MW (1975) A convenient correlation for heat transfer to constant and variable property fluids in turbulent pipe flow. Int. J. Heat Mass Transfer 18: 677-683.

So RMC, Sommer TP (1996) An explicit algebraic heat flux model for the temperature field. Int. J. Heat Mass Transfer 39: 455 – 465.

Sogin HH, Subramanian VS (1961) Local mass transfer from circular cylinders in cross flow. Journal of Heat Transfer 83:483 - 493.

Sommer TP (1994) Near wall modelling of turbulent heat transport in non buoyant and buoyant flows. Ph.D. thesis, Arizona State University, Tempe, AZ.

Spalding DB (1961) A single formula for the law of the wall. J. Appl. Mech. 28:455 - 458.

Sommerfeld A (1978) Vorlesungen über Theoretische Physik, Band VI, Partielle Differentialgleichungen der Physik, Verlag Harry Deutsch, Frankfurt.

Sparrow, Gregg (1959) Heat transfer from a rotating disk to fluids of any Prandtl number. Journal of Heat Transfer 81:249 – 251.

Speziale CG, Sarkar, S, Gatski TB (1991) Modeling the pressure-strain correlation of turbulence: an invariant dynamical approach. J. Fluid Mech. 227:245-272.

Speziale CG, Younis BA, Berger SA (2000) Analysis and modelling of turbulent flow in an axially rotating pipe. J. Fluid Mech. 407: 1-26.

Spurk JH (1987) Strömungslehre. Springer, Berlin Heidelberg New York.

Spurk JH (1992) Dimensionsanalyse in der Strömungslehre, Springer, Berlin Heidelberg New York.

Stein R (1966) Liquid metal heat transfer. Adv. in Heat Transfer, Academic Press, New York. Vol. 3:101- 174.

Stephenson (1986) Partial Differential Equations for Scientists and Engineers. Longman, London, UK.

Sternling, CV, Sleicher CA (1962) Asymptotic eigenfunctions for turbulent heat transfer in a pipe. J. Aerospace Sci. 29:109 - 111.

Taylor GI (1916) Conditions at the surface of a hot body exposed to the wind. British Advisory Commitee for Aeronautics, Vol. 2, R & M Nr. 272, pp. 423 – 429.

Taylor GI (1923) Stability of a viscous liquid contained between two rotating cylinders. Phil. Trans. Royal Society London A 223: 289 – 343.

Tietjins O (1970) Strömungslehre II, Springer, Berlin.

Törnig W (1979) Numerische Mathematik für Ingenieure und Physiker. Band 2: Eigenwertprobleme und numerische Methoden der Analysis. Springer, Berlin.

Vick B, Özisik MN, Bayazitoglu Y (1980) A method of analysis of low Peclet number thermal entry region problems with axial conduction. Letters in Heat Mass Transfer 7:235 - 248.

Vick B, Özisik MN (1981) An exaxt analysis of low Peclet number heat transfer in laminar flow with axial conduction. Letters in Heat Mass Transfer 8:1 - 10.

Weigand B, Beer H (1989a) Wärmeübertragung in einem axial rotierenden, durchströmten Rohr im Bereich des thermischen Einlaufs Teil 1: Einfluß der Rotation auf eine turbulente Strömung. Wärme- und Stoffübertragung 24:191 – 202.

Weigand B, Beer H (1989b) Wärmeübertragung in einem axial rotierenden, durchströmten Rohr im Bereich des thermischen Einlaufs Teil 2: Einfluß der Rotation auf eine laminare Strömung. Wärme- und Stoffübertragung 24: 273 - 278.

Weigand B, Beer H (1992a) Fluid flow and heat transfer in an axially rotating pipe: The rotational entrance. Rotating Machinery (Transport Phenomena), Kim JH, Yang WJ (eds), Hemisphere Publ. Corp., London, pp. 325 - 340.

Weigand B, Beer H (1992b) Fluid flow and heat transfer in an axially rotating pipe subjected to external convection. Int. J. Heat Mass Transfer 35:1803 – 1809.

Weigand B, Schmidt J, Beer H (1993) An analytic study of liquid solidification in low Peclet number forced flows inside a parallel plate channel concerning axial heat conduction. Proc. 4th Int. Symp. on Therm. Eng. Sci. for Cold Regions, Hanover, New Hampshire, USA, pp. 39 - 47.

Weigand B, Beer H (1994) On the universality of the velocity profiles of a turbulent flow in an axially rotating pipe, Appl. Sci. Res. 52:115 - 132.

Weigand B (1996) An exact analytical solution for the extended turbulent Graetz problem with Dirichlet wall boundary conditions for pipe and channel flows. Int. J. Heat Mass Transfer 39:1625 - 1637.

Weigand B, Ferguson JR, Crawford ME (1997a) An extended Kays and Crawford turbulent Prandtl number model. Int. J. Heat Mass Transfer 40: 4191-4196.

Weigand B, Wolf M, Beer H (1997b) Heat transfer in laminar and turbulent flows in the thermal entrance region of concentric annuli: Axial heat conduction effects in the fluid. Heat and Mass Transfer 33:67-80.

Weigand B, Kanzamar M, Beer H (2001) The extended Graetz problem with piecewise constant wall heat flux for pipe and channel flows. Int. J. Heat and Mass Transfer 44:3941-3952.

Weigand B, Schwartzkopff T, Sommer TP (2002) A numerical investigation of the heat transfer in a parallel plate channel with piecewise constant wall temperature boundary conditions. J. Heat Transfer 124:626-634.

Weigand B, Wrona F (2003) The extended Graetz problem with piecewise constant wall heat flux for laminar and turbulent flows inside concentric annuli. Heat and Mass Transfer 39:313-320.

Weigand B, Lauffer D (2004) The extended Graetz problem with piecewise constant wall temperature for pipe and channel flows. Submitted for publication to Int. J. Heat Mass Transfer.

White A (1964) Flow of fluid in an axially rotating pipe. J. of Mech. Eng. Sci.. 6: 47 - 52.

Worsoe-Schmidt PM (1967) Heat transfer in the thermal entrance region of circular tubes and annular passages with fully developed laminar flow. Int. J. Heat Mass Transfer 10:541 - 551.

Zagustin A, Zagustin K (1969) Analytical solution for turbulent flow in pipes. La Houille Blanche 2:113 - 118.

Zauderer E (1989) Partial differential equations of applied mathematics. John Wiley and sons, New York.

Zhang T, Osterkamp TE, Gosink JP (1991) A model for the thermal regime of Permafrost within the depth of annual temperature variations. Proc. 3rd Int. Symp. on Therm. Eng. Sci. for Cold Regions, Fairbanks, Alaska, USA, pp. 341 - 347.

Index

axial energy flow 123
axial heat conduction 47, 131
axially rotating pipe 70

Bessel equation 94
boundary conditions 20
boundary layer equations 192
   incompressible 174
   compressible 177
Buckingham theorem 181
bulk-temperature 57

canonical form 13
chain rule 13
channel flow 45
   laminar 58
   turbulent 60
Chapman-Rubesin parameter 194
characteristic equations 15
circular pipe 204
classification of second order partial
   differential equations 11
compressible flows 177
concentric annuli 159

d´Alembert solution 19
differential equation
   linear 9
   nonlinear, partial 6, 165
   partial 1
dimension matrix 182
dimensional analysis 180
direct numerical simulation 61, 210
Dirichlet wall boundary conditions 20
dissipation function 1
duct flow
   laminar 131
   turbulent 138
   fully developed flow 44

eddy diffusivity 89
eddy viscosity 206
eigenfunctions 53
   orthogonality 79
eigenvalue problem 11
   positve definite 51
   self-adjoint 51
   numerical method 227
eigenvalues
   asymptotic form 88
   large 87
energy equation 2
energy source 6

flat plate
   heated 174
flow index 63
forced convection 188
Fourier series 29
free convection 186
free variable 185
friction factor 64
fully developed flow 63

Graetz problem 43
   extended 117
Group-Theory 183

Hagen-Poiseuille flow 70
heat conduction 24
heat transfer coefficient 56
Hilbert space 124
hydraulic diameter 56
hydrodynamically fully developed
   channel flow 66
hyperbolic equation 12

Illingworth-Stewardson transformation
   193

incompressible flow 167
initial condition 20
inner product of the two vectors 124

Jacobian 13

Kirchhoff transformation 173

L'Hospital 244
laminar channel flow 136
laminar pipe flow 131
laminar sublayer 207
law of the Wall 210
liquid metals 117

matrix operator 123
mean axial velocity 206
mixing length model 45
    damping term 207

Navier-Stokes equations 1
Neumann conditions 20
normalization condition 78
Nusselt number 56

ordinary differential equations 17
orthogonal set of functions 54

parabolic problems 11
parabolic type 12
parallel plate channel 203
partial differential equation 1
    nonlinear 165
    elliptic type 12
    parabolic type 12
    hyperbolic type 12
Peclet number 48
Pi theorem 181
piecewise constant wall heat flux 157
piecewise constant wall temperature
    154

quadratic equation 11

relaminarization 222
Reynolds analogy 177
Reynolds number 58
Richardson number 74
rotating disk 165
rotation rate 70
Runge-Kutta method 228

semi-infinite body 2
separation of variables 24
shear stress 45
similar solutions 179
Similarity Solutions
    heat conduction 179
    boundary layer 186
speed of sound 6
standard form 12
stream function 193
stretched coordinate 94
Sturm-Liouville system 50
    properties 55
    standard form 90
superposition 7
symmetric operator 235

thermal entrance 43
total enthalpy 192
turbulent duct flow 138
turbulent Prandtl number 49

vector norm 243
velocity distribution 44
velocity potential 6
velocity profile
    fully developed 203

wave equation 18
wetted perimeter 56
WKB(J) method 87